T0220146

PRAISE FOR THE BOOK

Sonja Olson, an experienced emergency clinician, veterinary mentor, and health and well-being trainer of veterinary teams, shares her wisdom, her breadth of experience, and knowledge on the topics of well-being and resilience in this book. She also brings to life the voices of her colleagues across the globe. By doing so, she normalizes conversations around vulnerability and mental health. Sonja "talks the talk" and "walks the walk" without ego and will show how you, too, can have a long and fulfilling professional life as well as a personal life. This book brings to the reader both knowledge and experience that is useful to those studying and practicing in the veterinary profession. This will be a valuable addition to the libraries of veterinarians, veterinary technicians, nurses, and paraprofessionals.

Vanessa Rohlf, *BA (Hons), MCouns&PsychTh, PhD,*
Counselor, Psychotherapist, and Research Fellow

This is a great book for the vet community. It's chock full of models, resources, questions, practices, and personal anecdotes from those on the front lines. We all want to feel seen and heard, that someone understands us, and Sonja does a great job with that, normalizing the experience of the reader. I especially appreciated the understanding and compassion for what clients are experiencing, which might also help for us to understand why they sometimes say and do the things they do. Part textbook, part self-help, this is the ideal book for any veterinary professional.

Julie Squires, *Certified Compassion Fatigue Specialist*
and Certified Life Coach, Rekindle, LLC

Dr. Sonja Olson's passion is helping veterinarians thrive, and she is at the forefront of a grassroots movement to improve the well-being of both individual practitioners and the profession. *A Call to Life* takes a much-needed, comprehensive look at the unspoken codes and culture of veterinary education and practice. Using humor and compassion, Dr. Olson charts a path forward to new era of veterinary practice, one where veterinarians can flourish both at work and at home. This book is essential reading for every practitioner and student.

Karen R. Fine, DVM, *Author,*
Narrative Medicine in Veterinary Practice, October 2021

This book makes some big - and sometimes hard to hear - points without lecturing: instead, it feels like having a heartfelt, non-judgmental conversation. It recognizes that while a good deal of the onus is on us to change our perspective, there are outside forces beyond our control (forces identified and addressed with as much care and importance). Reading this book is an emotional rollercoaster, but it ends on an optimistic and positive note without being stifling or overwhelming. *A Call to Life* is something we need early on in vet school: it would have changed my experience so much.

Veterinary Surgeon and Veterinary Rehabilitation Practitioner,
BVM&S, MRCVS, Arizona, USA

Creating Wellbeing and Building Resilience in the Veterinary Profession

A Call to Life

Sonja A. Olson

CRC Press
Taylor & Francis Group
Boca Raton London New York

CRC Press is an imprint of the
Taylor & Francis Group, an **informa** business

First edition published 2022
by CRC Press
6000 Broken Sound Parkway NW, Suite 300, Boca Raton, FL 33487-2742

and by CRC Press
2 Park Square, Milton Park, Abingdon, Oxon, OX14 4RN

CRC Press is an imprint of Taylor & Francis Group, LLC

ISBN: 978-1-032-20335-5 (hbk)
ISBN: 978-0-367-41879-3 (pbk)
ISBN: 978-0-367-81676-6 (ebk)

DOI: 10.1201/9780367816766

Typeset in Times
by Deanta Global Publishing Services, Chennai, India

To my favorite human who happens to be my husband, Matt, and to all of the animals that have been a part of my life's journey. I am deeply grateful.

Contents

Foreword

When I was seven years old, we lived on a small hobby farm in a semi-rural area of Victoria, Australia. My parents had just purchased six Hereford calves that they planned to raise and then later sell at the stockyards.

I had purchased one, too, with my pocket money. Her name was Sarah.

As the cattle were settling into their new surroundings, I was eager to go and meet our new "pets". Unfortunately, my eagerness was not shared by the small herd, and they moved away whenever I approached.

I did not, however, give up. I eventually discovered that if I lay still long enough on the ground, they would come to me.

I still remember the smell of the grass, the warmth of the summer breeze on my skin, the sound the cattle made as they came over to sniff me, this strange creature lying on their patch of grass. Having these kinds of connections with animals led to feelings of peace and happiness.

As you read this, you may think of your own early connections with animals, and by reflecting on this, you might see how your decision to move into the veterinary profession may have been shaped by these very experiences.

For many, being a veterinary professional is not just what we do. It is part of who we are.

When things go wrong, when we start to question our career—whether we have what it takes, whether we belong here—it affects us to the core of our identity, it can lead us to despair.

The flip side of this is that, when things go right, when we find our work meaningful, and when it leads to alleviating suffering and healing and saving lives—oh, the joy! Our purpose is fulfilled, and our hearts are full.

Such is the paradox of veterinary work. How do I know? I used to be a veterinary nurse. I have experienced, firsthand, the ups and downs of the profession.

Fast forward to today, I am now a registered (licensed) counselor and psychotherapist. I provide support through counseling, therapy, and education to veterinary professionals, and I hear stories just like this.

As I reflect on my work as a veterinary nurse which was over 20 years ago, the idea that veterinary work could impact on wellbeing was not really acknowledged or discussed. Perhaps I was naïve, but I certainly hadn't heard of the concepts of burnout, compassion fatigue, and secondary traumatic stress. The work did, however, at times affect me. It saddened me. It shocked me. It distressed me, and it also frightened me. I didn't talk about it. I thought it was just me.

Perhaps back then, in the medical profession, the idea that emotions impact working lives was seen as unprofessional and unscientific. Yet, it is the emotional connection we feel with animals which drew us into the profession in the first place. Further, what we also now know is that recognizing and managing emotions in clinical practice has real value. The skill enables veterinary professionals to provide compassionate care to clients and patients, and it promotes professional longevity by fostering resilience.

This lack of acknowledgment of the emotional toll of veterinary work was widespread. We can even see it in the academic literature. Apart from a handful of studies addressing high rates of suicide in the veterinary profession (Blair & Hayes, 1982; Kinlen, 1983; Miller & Beaumont, 1995), it wasn't until the early 2000s that research efforts addressing the mental health issues faced by veterinary professionals significantly increased. More recently, there has been a shift toward investigating factors linked to job satisfaction, resilience, and wellbeing in the academic literature (Platt, Hawton, Simkin, & Mellanby, 2012). We now recognize how the work can be stressful as well as satisfying.

Still, I notice a chasm. It exists between what we know about mental health, wellbeing, and resilience in the academic literature and what those who work in the veterinary profession know, live, and breathe. And this is where Sonja Olson's book truly comes to life.

Sonja Olson, an experienced emergency clinician, veterinary mentor, and health and wellbeing trainer of veterinary teams, shares her wisdom and breadth of experience and knowledge on the topic of wellbeing and resilience in this book. She also brings to life the voices of her colleagues across the globe. By doing so, she normalizes conversations around vulnerability and mental health. Sonja "talks the talk" and "walks the walk" without ego and will show how you, too, can have a long and fulfilling professional life as well as a personal life.

This book brings to the reader knowledge and experience that is useful to those studying and practicing in the veterinary profession, making it a valuable addition to the libraries of veterinarians, veterinary technicians, nurses, and paraprofessionals. Its value lies in Olson's acknowledgment of the not uncommon imbalance in the journey of becoming a veterinary professional, with its emphasis on clinical skills, striving, precision, and doing in favor of self-care.

In the pursuit of this career, and even when we do attain it, we may focus on the next skill, hurry to the next client, or the next patient, not stopping to take a breath. We can very quickly lose sight of our seven-year-old self and forget why we are doing what we are doing in the first place, and we forget to be still— a sure recipe for burnout.

So, as you read this book, take it with an open and curious attitude, listen to the voices of those in your profession, do not give up, and know that you are not alone.

Dr Vanessa Rohlf, BA (Hons), MCouns & PsychTh, PhD
Counselor, Psychotherapist, and Research Fellow

REFERENCES

Blair, A., & Hayes, H. M., Jr. (1982). Mortality Patterns among US Veterinarians, 1947–1977: An Expanded Study. *International Journal of Epidemiology*, *11*(4), 391–397. doi:10.1093/ije/11.4.391

Kinlen, L. J. (1983). Mortality among British veterinary surgeons. *British Medical Journal (Clinical Research Ed.)*, *287*(6398), 1017–1019. doi:10.1136/bmj.287.6398.1017

Miller, J. M., & Beaumont, J. J. (1995). Suicide, Cancer, and Other Causes of Death among California Veterinarians, 1960–1992. *American Journal of Industrial Medicine, 27*(1), 37–49. doi:10.1002/ajim.4700270105

Platt, B., Hawton, K., Simkin, S., & Mellanby, R. (2012). Suicidal Behaviour and Psychosocial Problems in Veterinary Surgeons: A Systematic Review. *The International Journal for Research in Social and Genetic Epidemiology and Mental Health Services, 47*(2), 223–240. doi:10.1007/s00127-010-0328-6.

Author

Dr Sonja Olson grew up with her human and animal family members mostly in Maryland, but every summer included time with family in Wisconsin as well. She graduated from Virginia-Maryland Regional College of Veterinary Medicine with a focus on exotic animal medicine. Her professional path led to over 25 years of practicing clinical small animal and exotic emergency medicine in both private and corporate practice environments. The myriad of opportunities to teach, mentor, and lead that arose during those years were deeply fulfilling, but she gradually recognized that there was a real lack of veterinary wellbeing awareness and skills. This combination fostered Sonja's passion to better understand, and then support, the holistic health of veterinary caregivers. She now focuses her energy on being a health and wellbeing trainer for veterinary associates through her current work role, podcasts, writing, and collaborative efforts with other like-minded souls. Creating awareness, heightening knowledge, and building compassionate communities will serve as the foundation for a thriving future veterinary profession. Reenergizing for her comes from enjoying playtime with her husband and furry kids, dark chocolate, yoga, running, and deepening her vipassana meditation practices.

Illustrator

Dean Scott did four years of hard time at the University of California, Davis Veterinary Correctional Institute and was released on his own recognizance in 1993. Twenty-eight years of ongoing therapy later, he almost qualifies as a fully functional human being. His hours and days spent practicing veterinary medicine every week really interfere with his life. He is an author/illustrator, inflicting his humor on the unsuspecting populace with such high-brow fare as *The Veterinary School Survival Guides* (From the Back Row and Vet Med Spread), the dog breed parody book, *The Incomplete Dog Book: Nothing You Ever Wanted To Know About Dogs*, two volumes of cartoons titled *The Lighter Side of Veterinary Medicine*, two volumes of animal cartoons titled *Menagerie*, two children's books, *Cowabunga* and *Callie*, and a short story collection of fiction, *Something for Everyone*. His song parodies and veterinary skits can be found on YouTube under funnyvetdotcom.

Introduction

In the veterinary caregiving profession, we have got some "issues" we need to discuss. We have a long-standing culture of perfectionism, of judging (ourselves and others), and, as a result, a whole lot of shame and unhealthy attitudes. This stigma that we "should" be smart enough, tough enough, selfless enough to be great veterinarians and technicians right out of the gate from our academic training is ubiquitous. We expect it of ourselves because it was indoctrinated before and during our clinical training. It is then reinforced in the clinical environments that we graduate into as we start to practice. This is the "novice" professional experience that is well-known and written about in the human caregiving literature. However, in veterinary medicine, we are just starting to acknowledge and apply the concepts from the human medical research and literature on the "occupational hazards" of being in a caregiving profession. Such terms as "compassion fatigue", "vicarious trauma", and "moral stressors" are a few examples that we are just in the last few years becoming familiar with and applying to what we have been feeling and experiencing in this profession for many decades.

In my 25 years as an emergency clinician, veterinary mentor, and supporter of our veterinary teams on wellbeing, I have experienced the many highs and lows that our profession delivers. The incredible feeling of helping a suffering animal recover and of working like a "well-oiled machine" with my ER team as we provided high-quality patient care are the highs. Teaching and being taught daily in clinical practice were deeply fulfilling. Yet, like many others, I suffered from the empathic distress, secondary trauma, and burnout that can come from being a deeply empathetic caregiver. I personally experienced the fear and feelings of entrapment and the identity crisis of not knowing what to do if I was not "Dr Olson, the emergency clinician" in a clinical environment. Through deep work in self-reflection, honesty, and clarifying my values and intentions, I have found my way to a healthier space where I am contributing to our veterinary profession and community in a way that I would not have imagined ten years ago. I am honored to now be in a novel role that provides wellbeing support to the amazing veterinary professionals who continue to selflessly give in the veterinary practice environment. Do I miss being on the clinical floor and working with veterinary patients and fellow veterinary colleagues? You bet. Do I feel like I am intellectually less engaged or excited about my professional development? Absolutely not! I have given myself permission to show up as my authentic self and to serve our profession in a way that contributes to my holistic wellbeing and to those that I support.

Throughout those many years, shifts, and roles, I am saddened by the amount of human suffering I have witnessed in the name of selfless caregiving. The unhealthy coping mechanisms that individuals develop, the sacrifice of time for self-care and for family, and the number of individuals who "burn out" and leave this profession honestly break my heart. How many might have stayed if given the opportunity to understand how they might have more skilfully navigated the hurdles and pain that led to the decision to leave or, God forbid, to harm themselves?

My goal with this book is not to provide a formulaic recipe of how to avoid or to "fix" these issues. Why not? The honest answer is that this is a far more complex and individualistic journey. Each one of us has to determine the "when", the "what", and the "how" to support ourselves as we proceed through our lives. What worked for me may or may not jibe with you and your concerns. To that end, it is paramount that we each cultivate a sense of self-awareness, of self-worth, and of self-compassion. We must own our "stuff"—the good and the rough. It is our responsibility to educate ourselves and become more aware of the occupational hazards related to caregiving. We then have to dedicate ourselves to creating a set of tools and solutions that not only support our improved wellbeing but also contribute to the shift in our veterinary culture toward one of honesty, vulnerability, and compassion—for ourselves and each other!

If you are seeking clarity and the opportunity to build a more "quake-proof" foundation to build your work–life integration upon, this is where we start. Have the courage to look inward and embrace your imperfect human self with candor and with kindness. We are then better able to allow ourselves to celebrate the amazing, inspirational work that we do to better the lives of animals in all the ways that we do as veterinary professionals every day.

This is not a book prescribing solutions or treatment to situations or conditions. Although you will find I reference white papers, books, and research, the primary focus of this book is to create a living narrative told by amazing veterinary professionals in different seats and in different countries.

Whether I reference them directly or not, the experiences and wisdom of these veterinary colleagues are embedded in my perspective and the words I chose here. As much as I had wanted this book to be out in the world last summer, I genuinely believe that it is better now as a result of the knowledge I have gained and the conversations I have had with colleagues over the course of the last year. The personal and professional challenges associated with this year of the Covid-19 pandemic and all of the other unprecedented socioeconomic and political challenges in our respective communities have exhausted us, motivated us, and taught us. We are more ready than ever to have the long-overdue difficult conversations that will address dysfunctions in so many parts of our lives.

Now, I invite you to this honest, open conversation. I believe we can, and deserve to, have a fulfilling veterinary career, *and* have a life that we love. We are a community that is dedicated to being humane to all beings and to supporting health and longevity. Let's get started on fortifying ourselves and taking a stand for cultivating a professional culture of compassion and collaboration. We belong. We matter. We are more than capable. I believe in us.

> This is the secret of accompaniment: I will hold a mirror to you and show you your value, bear witness to your suffering and to your light. And over time, you will do the same for me, for within the relationship lies the promise of our shared dignity and the mutual encouragement needed to do the hard things.

> **—Jacqueline Novogratz,** *Manifesto for a Moral Revolution*

1 The Evolution of the Veterinary Profession
How Did We Get Here?

If you don't know where you are going, you'll end up someplace else.

—**Yogi Berra**

OUR SHARED HISTORY

The historic cultural expectations and ways of working in the veterinary field have created a system that is resulting in burned-out caregivers, who either leave the profession or disengage from their lives.

By examining the high cost to us as veterinary professionals, we can speak the truth of what is no longer serving us as a profession and collaborate on solutions to create a healthier culture and more professional systems. Ultimately, we need to have a deeper understanding of our profession and ourselves to work toward sustainable solutions that will allow us to experience more joy and thrive as caregiving individuals and professionals.

As such, we need to have a clear sense of how we arrived where we are today as a profession.

To set the stage for this book's story, I begin with a historical perspective of our complex veterinary profession culture. It is interesting and valuable to consider the evolution of the relationship between humans and animals in their care. Here, I present a high-level overview of the history and evolution of veterinary medicine to lay down a foundation. Although there are some veterinary fields of medicine that have maintained a more "traditional" relationship between caregivers and the animals in their care (think production animal medicine), there are other fields that have changed significantly, such as companion animal medicine. This has resulted in a different set of expectations of the veterinary caregiver from the owner/guardian of the animal.

Prior to delving further into these more specific and modern topics around the human–animal bond, I want to take a moment to step back and share the broader, historical view of the veterinary profession.

"Trivial Pursuit" time! Did you know that the word "veterinary" comes from the Latin *veterinae*, meaning "working animals", and that the word "veterinarian" was first used in print by Thomas Browne in 1646? Take a moment to "geek out" with me

DOI: 10.1201/9780367816766-1

and peruse the provided timeline which depicts our long-standing, global relationship with animals.

GLOBAL TIMELINE OF VETERINARY MEDICINE

BC

3000 BC	Mesopotamia. The Mesopotamian, Urlugaledinna, named as "expert in healing animals".
2500	China. Veterinary treatises on horses and cattle.
2200	Babylonia. Code of Hammurabi outlines veterinary fees.
1800	India. Salihotra named "veterinarian" of horses.
400	Greece. Hippocrates "humoral pathology" affects veterinary practice for 2000 years.
250	India. Kin Asoka constructs veterinary hospitals.
70	Rome. Varro treatise indicates a veterinary professional class.

AD

330	Byzantium. Apsyrtus called "father of veterinary medicine".
450	Rome. Vegetius books on veterinary art, influential for years.
900	England. Anglo-Saxon Leech Book includes animal cures.
1350	Italy. Laurence Rusius Hippiatria widely circulated in printed edition after 1530.
1490	Spain. Short-lived veterinary schools established.
1522	Spain. Francisco de la Reyna Book of Veterinary.
1528	Switzerland. Vegetius work printed as Mulo-Medicina.
1565	England. Thomas Blundeville first major English veterinary book on horses.
1576	England. George Turbeville first English book dealing with diseases of dogs.
1598	Italy. Carlo Ruini first anatomy of the horse, prefacing the start of veterinary science.
1639.	England. Thomas de Grey book on horses, hereditary disease, and attempted rationale for common procedures.
1664	France. Jacques de Solleysel wrote classic text, recognized glanders.
1683	Scotland. Andrew Snape first English equine anatomy book.
1711	Italy and England. Giovanni Lasci and Thomas Bates both establish effective methods to control rinderpest (but was not used).
1720	England. William Gibson, farrier, advances humane treatment, rational medication, and education.
1761	France. Claude Bourgelat founded Lyon Veterinary School and Alfort School in 1765, "start of the veterinary profession".
1783	England. Francis Clater. Every man his own farrier—the first of many horse doctor books.

1785	England. Oldham Agricultural Society proposes British Veterinary School.
1791	England. London Veterinary College founded with Sainbel as first professor, "start of the British veterinary profession".
Early 19th century	United States. First veterinary schools established in Boston, New York, and Philadelphia
1863	United States. American Veterinary Medical Association founded.
1879	Iowa, USA. Iowa Agricultural College became the first land grant school to establish a school of veterinary medicine.
1965	United States. FDA added a Veterinary Medical Branch to oversee veterinary pharmaceuticals (later became the Center for Veterinary Medicine).

MODERN DAY DEFINITION OF *VETERINARY MEDICINE* IN WIKIPEDIA (2020):

- Prevention, control, diagnosis, and treatment of disease, disorder, and injury in animals.
- Also deals with animal rearing, husbandry, breeding, and research on nutrition and product development.
- The wide scope of veterinary medicine covers all animal species, both domesticated and wild.
- Veterinary science helps human health through the monitoring and control of zoonotic diseases, food safety, and, indirectly, human applications from basic medical research.
- Veterinary medicine and science also maintain the human food supply through livestock health monitoring and treatment keep pets healthy and long-living.
- Veterinary scientists often collaborate with epidemiologists and other health and natural scientists, e.g., the global One Health initiative—a concept that describes the wellness of humans, animals, and the environment as permanently tied together.[1]

Veterinary medicine and the state of the profession directly correlates with the ever-evolving relationships that humans have with animals. Regarding domesticated animals, there were initially practical reasons for interest in their health. Animals have provided protection and service to humans for more than 15,000 years. These services range from providing food, supporting farming and hunting, and reducing the vermin populations in barns, which decreases transmission of disease and damage to food or other materials. A historic example is the horse, a primary focus of veterinary medical care over the last 2000 years, as they have been economically important for transportation, agriculture, and trade.

As human medicine and understanding of disease and pathology progressed, so did veterinary medicine. The concepts were applied to farm animals first in the

1700s. The anatomy and diseases of horses, cattle, and sheep were studied with great interest due to the animals' importance to the agrarian economy. Veterinarians were largely called upon to care for livestock—particularly horses—due to military and agricultural needs.

Over the last 100 years, there have been tremendous shifts in veterinary medicine. It was only in the early 1900s that domesticated animals, such as dogs, cats, and exotic pocket pets (e.g., rabbits and guinea pigs), came to be viewed more as pets and, as a result, received more regular medical care and improved nutrition. Particularly over the last 40 years, the small animal veterinary medicine community's focus expanded to include domesticated animals that were increasingly seen as family members, although they may still have their "jobs" of being guard or herding dogs or vermin-control cats. The advent of improved parasite control, both GI parasites and ectoparasites (such as fleas and ticks), supported this transition of animals from the yards and barns into our homes.

THE HUMAN–ANIMAL BOND OF TODAY

As the human–animal bond has evolved over the 20th and 21st centuries, so have opportunities and challenges for veterinary professionals. It is exciting to have so many more diagnostic and therapeutic options available for veterinary patients as well as an increased desire in many parts of the world to seek out and accept veterinary professional expertise and care. However, the flip side of this coin is that there are more fraught emotional obstacle courses that the veterinary caregivers must navigate with animal caretakers. Although your thinking may go immediately to a scenario that involves a much-loved "furry kid" in a client's home, there are other less obvious areas where emotions can run high, such as in zoo and wildlife conservation medicine. Recently, in some circles, the term "guardian" is used instead of "pet parent" to curb the anthropomorphizing of the relationship between animals and their caretakers and mitigate this confusing, emotionally charged landscape. All of this matters because it impacts how veterinary caregivers are perceived by the animal "guardians", how we perceive ourselves, how monies are invested (or not) in animal care, and the creation of laws to protect animal welfare and rights.

American Veterinary Medical Association Definition of the "Human–Animal Bond" is:

> The human–animal bond is a mutually beneficial and dynamic relationship between people and animals that is influenced by behaviors essential to the health and wellbeing of both. This includes, among other things, emotional, psychological, and physical interactions of people, animals, and the environment. The veterinarian's role in the human–animal bond is to maximize the potentials of this relationship between people and animals.
>
> The AVMA recognizes: (1) the existence of the human–animal bond and its importance to client and community health, (2) that the human–animal bond has existed for thousands of years, (3) that the human–animal bond has major significance for

veterinary medicine, because, as veterinary medicine serves society, it fulfills both human and animal needs.

Note: Human–Animal Bond Certification Program offered through the Human Animal Bond Research Institute (HABRI) and the North American Veterinary Community (NAVC).

There is no doubt that the way we care for veterinary patients, particularly those seen as "family members", has drastically changed over the last 30–40 years. Owners have more information (thank you, Dr Google!) and often wish to have a more active role in their pet's care. The thought process has also evolved from what would have been done for a working dog or a barn cat in the past to "What type of care would I want for a member of my family?" This humanizing of care for veterinary patients has allowed for remarkable advances in the types of care that can be provided today. More and more pets are considered part of the family. Some go so far as to refer to their pets as their "furry kids" and to themselves as "pet parents" rather than owners or guardians. We know that these evolving attitudes about animals and their status have also prompted changes in how we view wellness care, which now includes dental cleanings, grooming, and alternative treatments, such as naturopathic medicine and acupuncture. Services such as mobile veterinary medicine and hospice care have also grown from the desire to minimize stress and maximize care for both the veterinary patient and the animal guardian.

A poignant example of the evolving relationship of humans and their animal family members is when cancer is part of a differential diagnosis for a malady. According to the Animal Cancer Foundation's website (www.acfoundation.org), an estimated 65 million dogs and 32 million cats live in the United States. Of these, roughly 6 million new cancer diagnoses are made in both dogs and cats each year based on data from the National Cancer Institute of NIH in a 2017 article. The neoplasia that is being referenced may be either benign or malignant overgrowth of cells or tissues. Historically, a cancer diagnosis would have led to a conversation about palliative therapy and preparation for humane euthanasia to minimize suffering of the animal. Today, veterinary professionals have access to quite a few of the same diagnostics and therapeutics for the treatment of pets that, up until recently, were only used to treat people with cancer and other maladies. Importantly, not all owners are in a position to pursue these options, but the fact that we are at the stage in the human–pet relationship where many owners are asking to discuss these options, and may pursue them, is remarkable.

Certainly, ongoing research over the last several decades has demonstrated that human caregivers are benefiting from the relationship with animals as well. For example, people living with dogs are 15% less likely to die from heart disease, as stress levels are reduced, and caregivers enjoy increased activity levels from caring for and exercising the pet.[2] We have also learned that owning and caring for pets can lower blood pressure and raise blood oxytocin levels which are helpful for all humans—not just those with the potential for cardiomyopathies. Interacting with companion animals in elderly patients and with children can provide significant emotional, social, cognitive, and behavioral support. For all, owning a pet can also provide pet owners with a purpose which supports overall wellbeing.

There are people who do have a genuinely deep emotional attachment to their animals. The general population may not be as influenced by this outlook as we are. Are the animals benefiting as much from this bond as the humans do? If a medical concern is impacting the client, it becomes more likely that the pet will be brought in to see a veterinarian for evaluation and care in a timely manner. Think of the puppy with persistent diarrhea soiling the floors and furniture and wanting to go outside every 5 minutes, or the itchy dog that keeps his owners up all night with persistent scratching of his ears. Would that same pet be seen for comfort and appropriate medical therapy if it were not inconveniencing the pet owner? I recognize my own bias around this "have the pet seen now because it is bother-ing ME! (the client)", as I have helped many the pet and client during ER hours over the 20+ years that I was in clinical practice. However, I know that in speak-ing to colleagues in all veterinary segments then and now, this is a phenomenon that exists as part of veterinary medicine. I bring it up here as a situation that can evoke conflicting feelings for the veterinary care team. Is it really an emergency medical concern or a convenience visit for itchy ears at 3 am? The team is glad for the opportunity to care for the pet regardless of the hour, but why did the owner choose "now"? That last question can get emotionally "sticky" when it involves a more serious medical concern, such as a mammary mass that has been progres-sively enlarging for months but now has ruptured and is bleeding, and the mess in the house finally led the owner(s) to have the pet evaluated. Morally distressing for the care team? Absolutely.

There are other factors that may impact when, or if, a pet is seen regularly for wellness visits or is seen for a medical issue. Examples may include differences in perspective of the relationship that the pet has with the family, financial constraints, or beliefs in how much is "reasonable" to invest in the care of an animal, as well as variances in religious and cultural beliefs. In the veterinary caregivers' minds, there are pets that receive what many would consider the bare minimum of ethical care. By law, an animal is deemed to be "reasonably" cared for if food, water, and shelter are provided, even if each of these is questionable in quality and amount. It can be morally distressing to the veterinary professional to be the one that cares for the animal more than the owner appears to care. As veterinary caregivers, we are the advocates for the animal patients and their wellbeing. We will come back to this complex topic later in the book when we discuss ethical conflicts and moral stressors as contributors to veterinary caregiver burnout.

ONE CAREGIVING APPROACH (OR WAY OF THINKING) DOES NOT FIT ALL

As we examine the human–animal bond further, we also recognize that the relationship between humans and the animals that they care for can vary widely based on the role the animal may have in the human guardian's life. In conjunction, the veterinary caretaker's professional involvement in the animal's care can be, and will often be, associated with the "sector" of veterinary medicine within which they practice. Consider for a moment what you believe the relationship to be between the veterinary caregivers and the animal patients to be in the following veterinary environments: companion animal veterinary medicine, equine medicine (e.g., racehorses, show horses, hobby farms), food animal production (e.g., dairy, poultry, meat), military/police canine officers, zoo/wildlife conservation medicine, laboratory/research medicine.

Like you, I have my own opinions based on my life and professional experience to date. Additionally, in my years preceding veterinary school and in the 26 years since vet school graduation, I have had the pleasure to know many veterinary caregivers in each of these different veterinary caregiving positions. With those perspectives in mind, let's very briefly examine some of the unique relationships between animals and their guardians in different veterinary segments.

Companion Animal Medicine (Cats, Dogs, Birds, Small Mammals, Reptiles)

Clients in this veterinary segment are more likely to have an expectation for these animals to be treated like human family members and, resultantly, may have a high emotional attachment. However, this may or may not translate into a willingness to invest financially in the animal's care/wellbeing.

> **Dr Teresa Lightfoot—DABVP in Avian/Exotics is well-known as a clinician, instructor, speaker, and author around the care of exotic pets.** She stated that "after 35 years of practice and rationalizations that go along with that …" she had a few important takeaways.

The most important revelation I had was that pet owners just do not want to feel guilty. When presented with a recommended diagnostic and treatment plan that is way out of their budget, they feel awkward, guilty, and then, often, angry. Working into the conversation early, 'I know this might not be practical for you—you have a family to take care of' almost immediately removed the unspoken barrier between us (of financial concerns v being a good owner) and allowed us to work together to form a plan.

Dr Karen Fine, a small animal integrative medicine practitioner who has also practiced for several decades shared when it comes to supporting a family around financial constraints, she helps them assuage the guilt that may arise by sharing "that 'there is the ideal world and there is the real world'. I found myself saying something along those lines myself many, many times in the ER medicine realm".

EQUINE MEDICINE/SURGERY

The personal stories I have heard from equine practitioners offered contrasting perspectives on the pros/cons of this type of veterinary work. The patient care itself can be physically demanding, and, in many instances, the veterinarian may be seeing the patient by her/himself without technician assistance. The circumstances under which care is provided also vary depending upon whether the patient is out in a paddock, in the barn, or in a physical hospital offering more infrastructure and caregiving support to the veterinary team. It is not unusual to hear that some horse owners can be slow to seek veterinary care, as many remedies will be tried by the owners themselves first before reaching out to the vet. These owners can be quite suspicious of veterinarian care, and they may try therapies that the horse-owning community "knows to do" first and only contact the veterinarian if the medical issue does not resolve or worsens. The flip side of this approach can be seen with racehorses, which are very expensive and cared for at the highest level. These animals have veterinary caregivers who may be on call 24/7 or may even live on the premises to care for the animals. There are of course many, many horse owners that fall in between these two ends of the caregiving spectrum who work in tandem with their veterinary care team to ensure that proactive wellness and nutrition regimens are provided.

FOOD ANIMAL PRODUCTION

This is such an incredibly varied patient population—cattle, pigs, sheep, goats, poultry. The caregivers can have equally diverse relationships with these animals. The guardians of these animals may be less likely to have a close emotional bond to individual animals, but there is the prevalence of an ethical responsibility to have the DABVP animals lead healthy lives. Economics influence the preventative care and overall management of the herd (or flock). It also benefits the human consumer to have veterinarians involved in the design and implementation of healthcare protocols for these animals.

Dr Mike Endrizzi graduated from The Ohio State University in 1977 as a second-generation veterinarian in his family. He elected to practice mixed animal medicine for more than 17 years. He felt that when he graduated into his job and the veterinary profession, he and his professional peers felt they had "the best job". He enjoyed having work that provided case variety, intellectual challenges, the opportunity to build relationships with owners/farmers, and the ability to grow and learn throughout the years of practice.

MILITARY/POLICE DOGS

These special canine patients are cared for by both civilian and military veterinarians. A large amount of time and money is invested in the training of these animals. These dogs are often seen as protectors and partners and, as such, are highly respected and provided top care by their handlers and the others on the team.

In the years following my internship, I practiced small animal ER medicine in Alexandria, Virginia just outside of Washington, DC. It was an honor and a privilege to provide emergency care to the canine officers on the Alexandria police force as well as those from the Army base at Fort Belvoir. As an example, when the Pentagon was attacked on September 11, 2001, I was part of one of many veterinary teams that cared for the rescue dogs that day and for the week following. I witnessed firsthand the depth of the emotional relationship and the respect held for these dogs by their handlers.

ZOO/WILDLIFE MEDICINE

The patient populations of zoos, wildlife parks, and conservation management facilities certainly vary enormously as do their unique husbandry and medical concerns. What is not known or seen by the public are the often challenging and sometimes conflicting priorities between boards of directors, veterinary caregivers, keepers, and conservationists. If the animals are rare/endangered (e.g., snow leopards), or have been a part of the collection for a long time (e.g., gorillas), there is a very strong feeling of responsibility to do everything possible to care for these animals. Financial support can vary, too, depending upon the facility, the patient(s) being cared for, and the concern around public perception of the animal's wellbeing. This is a far more political and complex realm of veterinary medicine than those outside of it could possibly know.

In 1998, Disney's Animal Kingdom in Florida opened. Several papers and news outlets reported about the 31 animals that died at the park in the months before its opening. The USDA was involved in the investigation of the animal deaths. The federal report that was released said the animals died from accidents, poisonings (accidental toxicities due to plants and other substances found in the enclosures), fighting, and other forms of trauma (fence entanglements or injuries from other animals). I remember at the time hearing about this as well as from veterinary colleagues that worked at the National Zoo in Washington, DC, where I had volunteered for 10 years. It was enormously distressing to the keepers and the veterinarians to hear about the deaths of these animals; but even more so was how the Disney animal caregivers were being accused and treated by public and animal rights groups. Just an example of how the care and the death of these animals have been in the public eye long before the influence of the various social media that impact veterinary teams today.

LABORATORY/RESEARCH ANIMALS

Here again the patient population is quite varied, as are the environments under which care is provided. Over the last several decades, regulations, performance-based standards, and animal welfare have positively impacted research designs, protocols for care, and how animals can be used in research. Housing, nutrition, environmental enrichment, and consideration of animal welfare have all changed for the better as a result. The research done in these environments continues to play an enormously important role in furthering understanding of both human medicine, veterinary medicine, and One Health initiatives. While speaking to my colleague, Dr Dondrae Coble, he offered many insights from his numerous years in laboratory medicine, research, and teaching, such as the aforesaid improvements in lab animal care. When I inquired about the unique stressors for caregivers in these environments, he shared that work–life integration and creation of healthy workspaces for the professionals are as important in these environments as in any practice. Additionally, employee wellness programs are imperative in these environments, as concerns such as compassion fatigue are evident in these caregivers.

UNDERSTANDING OUR PAST AND CLARIFYING OUR PRESENT EXPERIENCES

In speaking to veterinary assistants, technicians, practice managers, and clinicians who practiced in the '80s and '90s, terms such as "compassion fatigue", "moral stressors", and "burnout" were not known or applied to the work-related experiences of veterinary professionals. With a current understanding of these terms, many of these individuals can appreciate, in retrospect, that they did in fact experience some of these conditions to a degree (or a lot!) when they were in practice. Those individuals that have been practicing since the late 1990s, including myself, have had the opportunity to witness firsthand changes in the veterinary landscape for the better and for the worse. As we examine the past, current, and future dynamics of veterinary practice and the impact on veterinary caregivers, there is wisdom in borrowing from existing learning from our human medicine colleagues. There exists abundant research and evolving strategies in human caregiving fields that can directly apply to veterinary medicine on how to practice high-quality medicine, maintain empathy and compassion, and not sacrifice mental and physical wellbeing along the way.

To wrap up the "How did we get here?" idea, meaning how veterinary medicine has changed in the last 30–40 years to what it is today with today's stressors and joys for veterinary professionals, let's hear from some veterinary colleagues. Each of these uniquely wonderful individuals shared how different things were for them when they first went out into clinical practice: different expectations of themselves, different client expectations, different relationships between the clients and the pets/animals, there was no social media/internet for a better part of their careers; and referral/specialty medicine was still primarily in the universities (not in private practice). These are their perspectives.

1. **Julia Jones (small animal general practitioner, Florida)**

    Each generation has challenges. Time organization has been my biggest challenge, but personal relationships have been my greatest source of reward. Our class of 1985 started out with the standard texts, classic journals, class notes, information gathered from colleagues and referral specialists, working crazy hours, sleeping at the clinic overnight or taking animals home on IVs. Emergency clinics were in their nascency and the nearest referral practices were often universities. Internet access was only for the professors. We now have access to more information at the touch of a finger, more specialists, and more emergency/ICU practices. One would think it improves time management. Standards of care have changed. Costs have changed. Legal challenges and regulations have changed. Veterinary-driven continuing education has shifted to industry. Sheer volume of available information has changed.

    With that comes the compartmentalization of veterinary care, as it has for human medicine. In our effort to "keep up", deliver care, see more patients, afford the cost of running a practice, and maintain our sanity, we have tried to restrict access to our "time" by more referrals. We are in danger of losing our personal relationships with the client if we are not careful. That "thing" that makes us special with people because of our love of animals and the compassion we show them is precious. This has been what sets us apart from the other health professions. My biggest advice to young veterinarians is to give more guidance to pet owners as they look to you for help in making decisions and not just throw a bunch of differentials and choices in their laps. Know the pet's name and the owner's name when you walk in the room. Listen and guide the conversation. Take interest in the outcome. Call them. It is amazing how much that binds them to you and the practice and increases the personal reward.

2. **Susan Hazel—senior lecturer, University of Adelaide, Australia**

    I graduated in 1987 when my experiences as a student and veterinarian were the same but different. As a graduate veterinarian, the challenges in moving from student to veterinarian were the same as for students today: not feeling that you knew enough, initially, but learning every day so the next case was easier. Feeling the full impact when animals had to be euthanased and owners comforted. The main difference is the internet—social media amplifying mistakes (and successes) and pressures of time and economics. I've moved to an academic role, and the students are amazing, but the pressures of working and studying seem worse. We had a few textbooks and lecture notes to study—now, lecture slides, papers, books, and the internet mean information is limitless. Veterinary care is still about maximising welfare for animals. We need to do this while protecting the welfare of veterinarians.

3. **Michael Schaer—practicing critical care medicine since the 1980s and still on faculty with UF Veterinary Teaching Hospital**

My impressions now span 51 years of veterinary medicine. I find that the profession has been blessed with many bright minds over the years, and I am glad to say that this trend has continued over the entire time period. I know some have stated that the new generation is spoiled and wants things handed to them on a platter, but I find that the majority of my students have consistently met the educational challenges, and that my graduates are better trained than ever before when they cross that stage to receive their diplomas.

I find that today's graduate is much more adept with the various technologies, but this has come, in some cases, at the expense of adhering to the basics when addressing the sick animal. They are the product of the education that is practiced in the teaching hospital. If they see clinicians jumping right into AFAST before they complete a detailed physical examination, then we are failing them right from the starting line by their failure to routinely do thorough abdominal palpations and other basic maneuvers, including rectal examinations and fundic examinations.

There are some schools that have increased their caseloads to the point of forcing some clinicians to the point of not spending enough personal time to take thorough histories which turn out to be very different than what is abstracted by a third party. I witness this to the present day, which ends up costing more time, money, and possibly harm. We should not abandon the teachings of Sir William Osler who taught that "Medicine begins with the patient, continues with the patient, and ends with the patient".

4. **Ken Drobatz—practicing ER and critical care medicine since the 1980s; clinical faculty & instructor at University of Pennsylvania Veterinary School & Teaching Hospital**

   i. Rapid and huge development in the sophistication of veterinary medicine.
   ii. The development of information technology and how it has sped up research and communication in veterinary medicine.

iii. The explosion of veterinary specialists and specialty hospitals.
iv. The massive increase in the costs of veterinary education and its impact on new veterinarians.
v. The massive increase in the cost of veterinary medicine to our clients.
vi. The initially gradual, but now rapidly increased, corporate ownership of vet hospitals.
vii. All of this could be contributing to the loss of the "soul" of veterinary medicine that it once had with its personal relationships and interactions with our clientele.

5. **Laura Buscher, DVM (and also known as "Nana" by her human family) practiced small animal medicine over the 33 years of her clinical career with a few farm animals sprinkled in "for entertainment". She owned her own practice as a solo doctor in a rural setting, was a relief veterinarian, a veterinary recruiter, and practiced in a large 50-doctor GP/specialty hospital before landing in her current role as a practitioner.**

How did we get here and where do we want to go? Winnie the Pooh says: "I always get where I am going by walking away from where I have been". We got here by grit. In a time when many said we couldn't, we simply kept doing. We "did", despite gender, marital status, or having children. To move forward we must knit ourselves together with our common passions and support each other's differences. I see women trying to define/confine women by a role that excludes other women. The more we include others, the better we will all be as individuals.

6. **Janis Massaro—technician at University of Florida Veterinary Teaching Hospital for 10+ years before moving into a veterinary practice manager role with an equine specialty practice for 17 years, now in clinician recruiting role for small animal ER/specialty practices (corporate group)**

Having been in the veterinary industry since 1991, I have seen a lot of changes over the years. When I started as a vet technician as head of Internal Medicine service at one of the nation's top vet schools, I had no idea this would lead me to 30 years in this amazing profession. When starting out in the nineties, we had no personal cell phones, no social media, and we (25 vet techs) worked closely as a team. We all had much respect for each other and would do anything to help our teammates. I attribute our success to our amazing vet tech manager who not only worked with us, she appreciated everything we did. My career took me to the management position (1998–2015). A day didn't pass without me telling my 35 technicians how amazing they were (this was the key for retention). In 2015, I became a recruiter. I now watch as social media, lack of support and appreciation, and exhaustion drive our veterinarians to depression and, likely, the need to leave their occupation from where they have dedicated most of their adult lives. I am hoping appreciation for all employees of the veterinary profession finds its way back. I truly believe thankfulness has become a thing of the past. It is what keeps us going. First and foremost, pay attention to the people you work with. Thank them and help them out every chance you get. Be visible. Be warm and tell them how important they are. In my opinion, this is the key to retention today, keeping our great profession safe and strong.

7. **Teresa Lightfoot, DABVP (exotics)**, who I quoted earlier in this chapter

    Forty years ago, the contract for my first associate veterinary job read "No more than 60 scheduled hours per week". However, that didn't include being on call every other week, for four different hospitals. This meant that almost every night I was back at the hospital multiple times for emergencies. No technical staff was available to help—emergencies were handled totally on your own. Work–life balance was not even a consideration. As my family grew, both they and I suffered from this. But it was the norm at the time. What we didn't have to contend with was social media. I sympathize with young veterinarians who have clients becoming armchair diagnosticians via Google and complaining publicly about cost. This must be disheartening and demoralizing. Forty years ago, we were granted more automatic respect. I hope that this text can assist these veterinarians in finding their way in the current professional climate.

INTENTIONALLY CREATING OUR FUTURE

"Progress is impossible without change, and those that cannot change their minds cannot change anything. If you change the way that you look at things, the things you look at change".

—George Bernard Shaw

The discussion at hand in the global veterinary profession is how to shift our community and professional environments effectively and positively toward sustainable, healthy dynamics. It is true that the responsibility to have a fulfilling life lies largely with ourselves. Only we can discern whether we are in a professional role where we feel engaged, supported, and valued. It is important to know yourself well enough to know whether the pace of work, the team that you are working with, and the community that you are serving suits you. Only you can know your values and whether the work you are doing is in alignment with these values and beliefs. Your understanding will evolve and clarify as your professional identity matures with clinical experience and life wisdom. Additionally, creating a life that includes self-care and replenishment must be thoughtfully and intentionally given time and energy by you.

Importantly, what is not your responsibility alone, however, is the systemic culture that you are working within. There needs to be just as much of a focus at the system and organizational levels—not just at the individual associate level—to create and support environments where veterinary medicine is done well and the veterinary caregivers are provided what they need to thrive. This is what I want to discuss next.

Before proceeding, take a moment to pause and reflect here. Ask yourself the following questions:

1. What type of medicine did you believe that you were going to practice and support while you were in school?
2. If that professional path changed for you, what were the factors that contributed to that shift?
3. How much of a role did finances play in the job that you accepted and may be staying in at this time?
4. Excluding finances as a factor, what would be your top three most important criteria for selecting your ideal job/team/role?
5. What else do you want to make room for in your life outside of being a veterinary professional?

In the many years that I have been in veterinary medicine in roles varying from intern, to ER clinician, team lead, mentor, clinician recruiter, and, now, wellness trainer, I have had an abundance of meaningful and candid conversations with veterinary technicians and clinicians regarding their state of satisfaction with their current professional and personal circumstances. We are going to take some time to dig into the specific concerns that are tied to the caregiving professions as well as identifying how these concepts may uniquely apply to veterinary medicine. It may be useful to take a closer look at some of the contributing factors that may lead to a more toxic versus a more fulfilling workplace. These factors may include the dynamics in professional relationships within the different environments where veterinary professionals care for animals, the implementation of available technology, the intentional development of a culture in classrooms and practices that support cultural humility, trauma-informed care, and a "people first" mindset to support veterinary professionals.

In this exploration, my hope is that there is a greater ability to have a more objective view of your role and of the professional environments that we find ourselves in today as veterinary caregivers. You are not the victim of your circumstances and have much greater autonomy and capacity to contribute to a positive, healthy work life, and environment than you believe. What we all have control over as of this moment is our individual mindset. With practice and ever-greater awareness, you are able to more effectively develop the capacity to reframe challenging circumstances. With clarity and intentionality, you can choose to see the difficult or uncomfortable moments as lessons uniquely tailored to help you to your next level of personal and professional development. There are no failures when you approach your life with this growth mindset. Rather, there are only opportunities for perspective and developing wisdom. My desire is to help you realize your own internal wisdom and to help you build and to fortify your wellbeing toolbox so that you can wisely respond to circumstances. It is absolutely within your capacity to identify and to then navigate challenging work-related scenarios, relationships, and conversations more skillfully.

Over the course of the following chapters, I invite you to join me on a journey of collective inquiry. We will explore what we have in common with other caregiving professions and how that can provide a helpful construct for more efficient solution-finding for our similar work-related concerns. We will also examine what concerns and considerations are unique to our veterinary caregiving profession and perhaps to the individuals that are attracted to the profession. Finally, I look forward to sharing some of the exciting and inspirational conversations and initiatives that have supported and are currently supporting the much-needed shift in how we treat ourselves and each other. I have had the honor and privilege of speaking with and reading the work of so many veterinary colleagues in preparation for the writing of this book over the last 2 years. Their voices will be heard through my interpretations as well as from them directly as quotes or references to their writings/podcasts.

I have been joking recently that I feel like a pregnant elephant that has been gestating this special baby, aka this book, for nearly 2 years. I have invested a lot of energy, time, and resources into this baby's development. Now, I am excited to deliver her into the world and share her with you. Let us all walk alongside one another and create a supportive, nurturing space for "this baby" to walk toward the bright and healthy future of our entire herd!

ROADMAP FOR THE JOURNEY AHEAD

Chapter 2 will review caregiving profession concerns, define terms, and provide context for these issues.

Chapter 3 looks at how the caregiving "occupational hazards" manifest in veterinary medicine and examines issues that are unique to the veterinary profession. In this chapter, for example, I will dig into some of the challenging personality traits to which many veterinary professionals are predisposed, how we might increase our awareness of them, and then purposefully and more skillfully practice veterinary medicine despite them. This is part of the larger conversation around self-compassion that includes three important components: self-kindness, common humanity, and mindfulness.

Chapter 4 is all about fortifying one's own toolbox of strategies for coping: increased self-confidence, control over one's destiny, and a sense of professional fulfillment with less personal suffering/sacrifice.

Chapters 5 is about a positive reframing of the discussions about where we are and where we want to go as individuals and as a professional community.

Chapter 6 is hopefully going to inspire you and provide you hope as we celebrate the unique and wonderful ways our colleagues are "showing up" to affect positive change in the veterinary profession.

NOTES

1. Takashima GK, and Day MJ. (2014). Setting the One Health Agenda and the Human-Companion Animal Bond. *Int. J. Environ. Res. Public Health*, 11, 11110–11120.
2. "The Human–Animal Bond throughout Time". College of Veterinary Medicine, Michigan State University, Perspectives Magazine, December 2018 (https://cvm.msu.edu/news/perspectives-magazine/perspectives-fall-2018).

2 Caregiver Concerns

> The highest activity a human being can attain is learning for understanding, because to understand is to be free.
>
> *—Baruch Spinoza*

The journey as a caregiving professional is an individual one for each of us, now and always. However, in order to prepare for a more robust, productive conversation about our individual and collective caregiving concerns, we need to be speaking "the same language" and be open-minded to the learning. There are terms and fundamental concepts that would be valuable to first clarify to lay the foundation for intentional and collaborative dialogue supporting us so that we can do what we love to do both as healthier individuals and as a healthier community.

There is wisdom and efficiency in considering what research, discussions, and solution-finding have already occurred in other healthcare professions outside of veterinary medicine. There are common themes that have led to terms and approaches that have been used in these other settings in various ways over the last 40–50 years that we are just now starting to apply to the veterinary profession. In this chapter, I will provide a high-level overview of some of these common caregiving considerations. It starts by clarifying and understanding some of the vocabulary/terms that are more commonly used in human healthcare to describe the "occupational hazards" of being compassionate caregivers to individuals that are suffering. This helps to set the stage to normalize the conversation and then more effectively collaborate in seeking solutions to those issues that also exist in our veterinary profession. There is no need to recreate the wheel. There is such an abundance of existing research and forward-thinking conversation supporting both the caregivers and patient care.

"It is the importance of the intense giving, and the fragility of it …" that has led to the effort of many researchers from different human medicine and psychology fields to evaluate how to support the holistic wellbeing of caregiving professionals.[1] When referring to the caregiving professions, this applies to all those that give of themselves to others whether the "other" be human, animal, or the environment. The care may be in the form of medicine, counseling, religious leadership, community activism, philanthropy, conservation/environmental work, or teaching. In all these realms, there exist common "work ingredients". Particularly since the 1970s, there has been a significant amount of research, data collection, and novel thinking supporting the exploration of areas that would be relevant across different disciplines and cultures. Three questions that embody the overarching goals of both historical and current exploration into caregiver thriving are:

- How do we foster flourishing and compassion satisfaction and prevent burnout?
- How does one balance between self-care and "other" care?
- What are the keys to the development of resiliency?

DOI: 10.1201/9780367816766-2

These are clearly invaluable questions. If easy answers existed, I promise that I would share them right away. I have a deep commitment to mitigating the suffering that results from a lack of awareness and understanding of common caregiving professional challenges. With that, join me in having a "beginner's mind" of curiosity and venture into helpful lessons learned to date from other caregiving professions.

VOCABULARY

RUMINATION (VS. REFLECTION)

Definition: Rumination is *the process of carefully thinking something over, pondering it, or meditating on it.* In psychology, the term refers to obsessive repetition of thoughts or excessively thinking about problems. Repetitively going over a thought or problem without completion.

Rumination is the focused attention on the symptoms of one's distress and on its possible causes and consequences, as opposed to its solutions. Both *rumination* and *worry* are associated with anxiety and other negative emotional states; however, their measures have not been unified.

Examples:

- "If only I had seen (fill in the blank) sooner, I could have (fill in the blank) and perhaps my patient would have lived!"
- "Why did I say (that) instead of (this)?"

Rumination is like an eddy in a river where you spend a lot of time and energy fighting the physical forces rather than tapping into your embodied wisdom and more skillfully navigating the challenging situation. Rather than rumination, if one opts to reflect on a situation with an objective curiosity and the intention to learn, it can lead to more impactful critical thinking and purposeful action. You can more clearly see the way the currents move, you have a broader perspective and, lo and behold, you paddle your way out of that swirling water back into the flow of the river.

EMOTIONAL LABOR

Definition: The process of managing feelings and expressions to fulfill the emotional requirements of a job.

Workers are expected to regulate their emotions during interactions with customers, coworkers, and superiors. The term "emotional labor" was first coined by sociologist Arlie Hochschild 37 years ago in the 1983 book, *The Managed Heart.* Emotional labor was used to describe the managing of one's emotions and affect at work to suit the expectations of the job. However, the term has evolved to also include unpaid, invisible labor or unrecognized work. Consider the unacknowledged yet expected requirement for veterinary professionals to maintain an aspect of friendly compassion, particularly toward clients, regardless of the circumstances or how we are being personally treated. It may feel like there is an unwritten rule that we are to "keep our cool" amid stress and high emotions that are not uncommon in the veterinary workplace, particularly with emergent patients. The motivation for such labor falls somewhere between caring and feeling like "we have to".

When we discuss "professionalism" in veterinary medicine, we may take that to mean that it is necessary to maintain a calm demeanor and stay in control of our emotions in the face of complex and often challenging circumstances. The problem here is that having to mask the inner turmoil to present the expected emotions can be detrimental to an individual's wellbeing. If our actual emotional responses do not align with the organization's "feeling rules", the individual then has to do additional work to manage their emotions. The psychological effort of this acting can be an important contributor to emotional burnout over time. If left unnoticed, or if not addressed, the costs can become toxic to that individual overall and result in physiological and mental health concerns. This is the cost of "cognitive dissonance".

Emotional labor is not just in our workplace. There are examples from our personal lives as well as work lives that include societal expectations of our "appropriate" behavior. In Adam Grant's podcast, "WorkLife", there is a terrific episode titled "Faking Your Emotions". In this episode, he interviews Alicia Grandey, an industrial-organization psychologist at Penn State University who studies how people manage emotions. She shares excellent insights into some of the pervasive and irrational emotional expectations of our larger society. Some of these societal assumptions might include: caregivers should "put on a happy face …" even when they are distressed or uncomfortable; women should subdue their emotions for fear of being seen as aggressive, weak, "crazy", combative, or irrational; men should not demonstrate vulnerability or sadness as this may be perceived as weakness in character or overall "toughness"; teammates showing a "brave face" when feelings have been hurt or someone's been insulted (e.g., racial slur or other prejudice) to demonstrate that they can "get along" or are not "too sensitive".

HEALTHY PRACTICES TO CONSIDER
AROUND EMOTIONAL LABOR

- Validate "heart work".
- Identify what emotions are expected to be managed as part of employment. What situations are likely to generate intensified emotional labor?
- Provide appropriate avenues for safe expression of organizationally inappropriate emotions after difficult situations.
- Protective factor of debriefing or even "deep acting" (where you play out a situation further) can recharge people after difficult workplace incidents if they are processing their feelings and tapping into their creative energy.
- Providing individuals with some time, private space, or access to a trusted colleague or an appropriate professional (e.g., veterinary social worker) will help them to "let it go".

STRESS

Definition: a state of mental or emotional strain or tension resulting from adverse or very demanding circumstances. It is simply the body's response to changes that create taxing demands. Today, we often use the word "stress" in everyday language to describe negative situations or feelings.

William Shakespeare in his poetic wisdom shared that "there is nothing good or bad, but thinking makes it so ..." It is our perception of circumstances as "stressful" that then causes us to feel that situation as being averse to our wellbeing or "bad" for us. We would be wise to be wary of overmedicalization of distress, which is part of the normal human experience.

The first person to study stress scientifically was a physiologist named Hans Selye. Dr Selye spent many years studying the physical reactions of animals to injury and disease. Based on his research, Selye concluded that human beings and animals share a specific and consistent pattern of physiological responses to illness or injury. These changes represent our body's attempt to cope with the demands imposed by the illness or injury process.

Richard S. Lazarus, PhD, (1922–2002) was professor emeritus of psychology at the University of California, Berkeley and was named one of the most influential psychologists in the field by American Psychologist. He was a pioneer in the study of emotion and stress in their relation to cognition. According to Lazarus, the effects of stress on a person are based more on that person's feelings of threat, vulnerability, and ability to cope than on the stressful event itself.

Embodied Symptoms of Stress

Physiologic responses to stress that we are familiar with are: the "fight or flight" response when our sympathetic nervous system is activated to protect us and the "fawn" response (parallels the "freeze" response to the stress). The physical symptoms that occur when we perceive a situation as a potential threat include increased heart rate, higher blood pressure, vasodilation, pupil dilation, bronchodilation, increased perfusion of our muscles with release of more fat/glucose for energy from the muscular tissue, slowed digestion (blood flow is shifted to muscles and vital organs).

Research has demonstrated that if the body is permitted to respond and then to return to homeostasis, it takes approximately 90 seconds to move through the four stages of the stress response.

Stage 1: Stimuli from one or more of the five senses are sent to the brain.
Stage 2: The brain deciphers the stimulus as either a threat or a non-threat.
Stage 3: The body stays activated or aroused until the threat is over.
Stage 4: The body returns to a state of homeostasis, a stage of physiological calmness, once the threat is gone.

Chronic stress response is when the stress is not allowed to "downregulate" and/or is persistent in an environment for the individual. This increased "allostatic load" can lead to GI disturbances, elevated cholesterol, hypertension, compromised immune system, increased anxiety, increased prevalence of clinical depression, insomnia, decreased libido, and decreased emotional regulation which can lead to impatience and increased irritability.

EUSTRESS (VS. DISTRESS)

Definition: Eustress is a term that refers to positive stress (appropriately received and responded to) vs. distress which refers to negative stress (perceived as and/or experienced as detrimental).

Eustress has the following characteristics:

* It is short-term.
* Motivates and focuses energy.
* Feels exciting (in a positive way!).
* Improves performance.
* Perceived as being within our coping abilities.

These characteristics make sense when you look at the overall natural arc of our "stress response". If there is a stress response that is perceived and coped with in a healthy way, our bodies and brains move toward what is called the "relaxation response" in about 90 seconds. This relaxation response has the following positive attributes: heart rate slows, decreased blood pressure, slowed brain waves, digestion improves as blood flow is higher to the gastrointestinal tract decreased blood lactate levels, immune system is strengthened, improved sleep, higher libido, and an improved overall sense of wellbeing.

Health Benefits of Experiencing "Eustress " on Physiology

* Oxytocin released during stressful events bonds us to our community and is an anti-inflammatory mediator to the myocardium.
* Increased heart rate—can be the same as when we are excited or are experiencing joy.
* Increased respiratory rate—invigoration of the body and the brain with increased oxygenation.
* Increased mental alertness if used in a mindful manner can allow us to be present and attentive to all that is occurring in that moment.

Consequences of Distress on Physiology

* Causes anxiety or concern.
* Duration can be short- or long-term.
* Feels unpleasant.
* Decreases performance.
* Can lead to mental and physical problems.
* Perceived as outside of our coping abilities.

Different people have different reactions to particular situations for a wide variety of reasons. We are complex individuals that each have our own unique physical, mental, cultural, and experiential filters through which we perceive stress. Thus, what may cause me to be distressed may not be a big deal in your book. Holiday seasons are a great example of events that may cause one individual joy and connection but may be a trigger of past traumatic experiences or a reminder of feelings of isolation for another. Keeping in mind that these are generalizations, here are some examples of positive personal stressors and negative personal stressors:

- *Positive:* starting a new job, getting married, having a child, buying a home, moving, taking a vacation, holiday seasons, retiring.
- *Negative:* separation from a spouse or committed partner, divorce, death of a family member or pet, hospitalization (yourself or someone close to you), injury or illness (for you, a family member, or a friend), financial concerns, unemployment, being abused or neglected.

These are intended to just provide some examples so that you can discern what is meant by "positive" vs. "negative" stressors. With that in mind, there is also a further differentiation that is important to mention when considering distress. There are stressors that may come from external sources (such as work) but there are also internal events that can be a source of negative stress.

Work and employment concerns, such as those listed below, are also frequent causes of distress:

- Excessive job demands.
- Job insecurity.
- Conflicts with teammates and supervisors.
- Inadequate authority necessary to carry out tasks.
- Lack of training necessary to do the job.
- Making presentations in front of colleagues or clients.
- Unproductive and time-consuming meetings.
- Commuting and travel schedules.

Stressors are not always limited to situations where some external situation is creating a problem. Internal events, such as *feelings* and *thoughts* and *habitual behaviors*, can also cause negative stress.

Common "internally caused" sources of distress include:

- Fears: (e.g., fear of flying, heights, public speaking, chatting with strangers at a party).
- Repetitive thought patterns.
- Worrying about future events (e.g., waiting for medical test results or job restructuring).
- Unrealistic, perfectionist expectations.

Habitual behavior patterns that can lead to distress include:

- Overscheduling.
- Failing to be assertive.
- Procrastination and/or failing to plan ahead.

Psychological Distress and Suicide (or Suicidal Ideation)

Psychological distress can manifest in multiple ways and at different levels of severity. In very general terms, it is psychological discomfort. It can be experienced as sadness, anxiety, distraction, and—in the most extreme cases—psychotic symptoms. It can be caused by many things: a severe one-time stressor, everyday or recurrent stressors, medical illness, or mental illness. It is a sense of discomfort and feeling unsettled—and usually at a level that is getting in the way of activities of daily living (e.g., work, school, caregiving, self-care). In less severe cases, psychological distress can be managed through rest, taking a break, and self-care, such as exercise. However, if the distress and the symptoms are really interfering with life, or leading to thoughts of harming self or others, compassionate intervention is required as quickly as possible, ideally with a licensed medical or mental health professional.

Symptoms of psychological distress may include behavioral problems, increased substance abuse, sleep disruption, poor work performance, feelings of worthlessness, chronic sadness, and inability to interact with other people. Psychological distress is measured based on the severity and length of the symptoms. Assessments often rely on an individual's self-reporting to mental health professionals. For instance, an individual might be asked whether they have experienced feelings of worthlessness, whether these feelings were fleeting or lasted awhile, and whether the feelings were manageable or unbearable.

The level of disruption to normal daily life is a major consideration when evaluating levels of psychological distress. A person's ability to work productively, eat a healthy diet, get a restful night's sleep, enjoy normal activities, and socialize are all considered when diagnosing and measuring distress. Of course, any suicidal thoughts or thoughts about harming others are always considered to be clear indicators of psychological distress.

Work-Related Traumatic Stress

Psychological Distress Types That Have Been Documented in Both Human and Veterinary Caregivers

- **Primary traumatic stress**: You are the direct target of a traumatic event. This can also refer to the psychological health impacts if you are the one experiencing a distressing situation. Being bitten by a suffering, frightened canine patient is a clear example of direct traumatic stress. Less obvious is the example of you being the one to witness a dog being hit by a car.
- **Secondary traumatic stress** (also called "vicarious traumatization"): Exposure to a traumatic event due to a relationship with the primary person (or animal). It is the witnessing or listening to the distress of a traumatized

person or animal. An example of this would be the impacts on you as a result of talking to a professional colleague who goes into detail about a critical patient that clinically decompensated, and CPR was unsuccessful when the patient arrested resulting in the animal's death. Other important callouts include:

a. Secondary trauma can happen suddenly. The emotional duress that results from hearing about firsthand trauma or witnessing the pain of another. This trauma type is more focused on behavioral symptoms (Note: The symptoms can mirror those of post-traumatic symptoms).[2]

b. Vicarious trauma can be a response to an accumulation of exposure to the pain of others (compassion stress). This trauma type is more focused on the cognitive response of the listener and its resulting impacts.

Symptoms of these two trauma types are nearly identical: emotional numbing, social withdrawal, work-related nightmares, feelings of despair and hopelessness, more negative view of the world, reduced sense of respect for your clients, reduced motivation, and increased illness.

COMPASSION FATIGUE VS. EMPATHIC DISTRESS

COMPASSION FATIGUE

Definition: The over identification with another individual's emotional pain.

Although the term "compassion fatigue" (CF) has come to be heard more frequently in veterinary circles over the last several years, it has been considered in other professional arenas for the last several decades. It was in 1995 that Dr Charles Figley published his seminal work, *Compassion Fatigue: Coping with Secondary Traumatic Stress*.[3] It was in this work that CF was first described as resulting "from stress when working in traumatic situations or witnessing others' distress". The symptoms that are associated with CF are: lack of empathy, anxiety, irritability, persistent arousal, dread of work, and avoidance behavior. Emotional, social, and spiritual exhaustion that decrease the ability to care for others.

Dr Figley's initial work around compassion fatigue was supported by that of Drs Saakvitne and Pearlmann who contributed to his 1995 book specifically on treating therapists with vicarious traumatization and secondary traumatic stress disorders. In their book, *Transforming the Pain*,[4] the authors suggest that the symptoms associated with CF and VT be looked at on three levels: physical, behavioral, and psychological. Everyone's symptoms will be unique to them, and evaluation of these signs/symptoms are not intended to be a diagnostic test. Rather, this examination is how we increase our individual awareness of how we react to the caring work that we chose to do as a profession. The opportunity to notice one's unique susceptibilities and how caregiving work may impact one's health can provide an "early warning system" and motivate self-care strategies. In her book *The Compassion Fatigue Workbook*, Francoise Mathieu provides an excellent summary of these symptoms and provides exercises for reflection and strategy building.[5]

Later work by Dr Figley and others on the "Compassion Fatigue Awareness Project" at Tulane University included veterinary professionals in their evaluations. In 2018, Dr Figley stated that:

Compassion fatigue is a state experienced by those helping people or animals in distress; it is an extreme state of tension and preoccupation with the suffering of those being helped to the degree that it can create a secondary traumatic stress for the helper (www.compassionfatigue.org).

EMPATHIC DISTRESS

Definition: The strong aversive and self-oriented response to the suffering of others, accompanied by the desire to withdraw from a situation in order to protect oneself from excessive negative feelings.[6] The concern with too much emotional empathy is that it can turn into this state of empathic distress.

When experiencing this distress, there may be a desire to withdraw from the situation or circumstances as a protective strategy guarding against too many negative emotions and being too upset. Remaining in this state of emotional overwhelm can have the same negative mental and physical effects that were identified with compassion fatigue. Interestingly, if empathy as "self-centered" is flipped to compassion which is "other-centered", it becomes a means to develop healthier emotional regulation and protective physiologic and mental states when in potentially distressing caregiving scenarios.

Stick with me here, as this is really important for us to understand. The words "compassion" and "empathy" are frequently seen as synonymous, but that is not accurate. There has been quite a bit of recent social neuroscience and behavioral science research that demonstrates very significant differences between these two mental constructs. An example is the work of Dr Tania Singer of the Max Planck Institute for Human Cognitive and Brain Sciences in Germany which shows that compassion fatigue is a misnomer and that *it is empathy that fatigues caregivers, not compassion!*[7]

Research has demonstrated that different areas of the brain are activated, and different associated neuropeptides are released, when an individual experiences empathy versus compassion. Empathy increases activation of areas of the brain that are associated with processing pain or threat, such as the amygdala. When these areas are chronically activated, dopamine levels are depleted. This can result in a blunted sense of reward and of pleasure. In healthcare professionals, chronic dopamine depletion as a result of empathic distress contributes to "emotional exhaustion, withdrawal, depersonalization, and a decreased state of personal accomplishment due to work-related stress".[8] There is no doubt that chronic empathic distress contributes to work-related burnout.

The antidote? *Cultivating compassion practices!* Compassion is the experience of feeling *with* the other versus empathy which is the feeling *for* the other. Research into the neuroplasticity of adult brains has demonstrated that compassion is a skill that can be practiced and developed. When practicing compassion, the areas of the brain that are activated are linked to the reward processing areas that then increase oxytocin and vasopressin (the bonding neuropeptides) as well as dopamine. Rather than feeling fatigued, expressing and experiencing compassion rather than empathy is neurologically and physiologically rejuvenating! Existing empathic distress can be reversed, and future distress mitigated, with the utilization of both self-compassion and of compassionate care for others.

EMPATHIC VS. EMPATHETIC?

Note: The words **empathetic** and **empathic** mean the same thing. **Empathic** is the older word, but not by much—it was first used in 1909, while the first recorded of use of **empathetic** is from 1932. Both words are derived from *empathy*, and you can use them interchangeably. In scientific writing, **empathic** is more common.

BURNOUT

Definition: A state of physical and emotional exhaustion that may occur in long-term high stress situations such as physically and emotionally challenging roles.

Burnout was first used in 1979 by Charles Maslach to describe the physical and emotional exhaustion that he observed in certain populations of caregivers. At that time, he was applying a term used to describe individuals who were deeply impacted by chronic illicit drug use. Chronic job-related stressors that are associated with low job satisfaction and fulfillment have been demonstrated to significantly contribute to an individual's experience of feeling "burned-out". Other factors include toxic relationships, incivility, and emotional and cognitive detachment.[9,10] Burnout may lead to a loss of desire to do the work itself because of unrealistic job demands and/or lack of support (both in terms of resources as well as psychological support). Trauma does not have to be present for burnout to occur but can exacerbate burnout development and symptoms experienced by an individual caregiver. Burnout does not apply only to those in the caregiving profession—it can affect any professional.

It is important to differentiate *burnout* from *compassion fatigue, secondary traumatic stress*, and *empathic distress*. Burnout is a work-related hopelessness and often results in feelings of inefficacy and dissatisfaction. Burnout is being "worn out" vs. secondary traumatic stress, which is about being afraid for your own well-being. Secondary traumatic stress can result from work-related secondary exposure to extremely or traumatically stressful events. The trauma did not directly happen to you. You were not in personal danger but were bearing witness to trauma whether auditory, visual, or both.

The physical symptoms of burnout certainly overlap with compassion fatigue, empathic distress, and with vicarious traumatization (secondary trauma). There can be feelings of being physically and emotionally depleted. There is often low job satisfaction because of inadequate time, resources (includes support staff), and autonomy in decision-making. These conditions can lead to a feeling of powerlessness and of doing inferior quality work.

Nothing gets to the heart of a medical caregiver faster than feeling that they are not able to provide the best care to their patient due to circumstantial reasons. Over time, if these systemic conditions in the workplace do not improve for the caregiver, feelings of helplessness, cynicism, and resentfulness can develop. While there are environmental and systemic stressors and conditions to consider with burnout, the individual caregiver must also examine their coping and adaptive skills. The combination of work-related causes, lifestyle causes, and personality traits can add up and compound each other to result in full-on burnout for that occupational role in that work environment.

In reviewing a classic Scientific American article, written in 2006, a 12-stage model of burnout developed by psychologists Herbert Freudenberger and Gail North was found.[11]

THE 12 PHASES OF BURNOUT

1. The compulsion to prove oneself: demonstrating worth obsessively; tends to hit the "best" employees, those with enthusiasm who accept responsibility readily.
2. Working harder: an inability to switch off.
3. Neglecting your needs: erratic sleeping, disrupted eating, lack of social interaction.
4. Displacement of conflicts: problems are dismissed; might feel threatened, panicky, and jittery.
5. Revision of values: values are skewed; friends and family are dismissed; hobbies are seen as irrelevant. Work is the only focus.
6. Denial of emerging problems: intolerance overall; perceiving collaborators as "stupid, lazy, demanding, or undisciplined"; making social contacts is harder; cynicism, aggression behaviors; problems are viewed as caused by time pressure and work, not because of life changes.

7. Withdrawal: social life becomes low priority or is non-existent; perhaps turning to alcohol/drugs; need to feel relief from stress.
8. Obvious behavior changes: odd behavior noted by friends, work colleagues, and family.
9. Depersonalization: seeing neither self nor others as valuable; no longer able to perceive own needs.
10. Inner emptiness: feeling empty inside; may turn to activities such as over-eating, sex, alcohol, or drugs to feel "alive".
11. Depression: feeling lost and unsure; holistically exhausted; future appears hopeless and dark.
12. Burnout syndrome: can include total mental and physical collapse; time for full medical attention.

The above symptoms clearly range from mild-but-worrisome to much more severe and in need of medical intervention. The 12 steps also do not necessarily follow one another in order and also take time to gradually develop. The key here is to be aware of these symptoms and perhaps be better prepared to pay attention to them sooner in yourself, or in a work colleague. Importantly, these symptoms are not necessarily an unavoidable part of a hard-charging, caregiving professional life.

Burnout as a Concern in Today's World and Steps Toward Relief

In 2019, the World Health Organization classified burnout as a defined clinical "syndrome" that results from "chronic workplace stress that has not been successfully managed". There is now a specific WHO International Disease Classification (ICD-11) in the official compendium of diseases. Burnout is an occupational phenomenon and not a medical condition. Burnout overall in all professions is when there is a mismatch between the workload and the resources needed to do the work in a meaningful way.

Possible clinical symptoms noted include:

• Feelings of energy depletion or exhaustion.
• Increased mental distance or feelings of negativism or cynicism related to one's job.
• Reduced professional efficacy.
• Limited to work environments and should not be applied to other areas of a person's life.
• Need to rule out adjustment disorder, anxiety, and mood disorders.

I appreciate that in both human medicine and in veterinary medicine, there is a new focus on factors that are causing burnout at the system level, not just at the individual caregiver level. There are interdisciplinary conversations occurring in human medicine through the AMA. As stated by Dr Christine Sinsky, MD, VP of professional satisfaction at the AMA, her committee work proposed that the best response to burnout is "a focus on fixing the workplace rather than focusing on fixing the worker".[12]

The key issue is to recognize that there are environmental factors that need to be addressed and that it is not solely the responsibility of the individual worker. The AMA's work around burnout as a concern has led to the creation of the "AMA STEPS Forward". There are 40 online modules which include resources, case studies, and other supportive content around patient care, workflow, leading change, professional wellbeing technology, and finance.

The recognition that it takes organizational leadership to make the necessary changes in addition to the team-based approach of "relieving physicians of tasks that drain joy from practice" is an enormous step in the direction of decreasing burnout for caregivers and increasing their compassion satisfaction. Importantly, these conversations are now being considered as part of the larger work in "organizational excellence".

I became aware of the *Shingo Model* as a powerful framework for organizational excellence during a panel discussion on veterinary burnout and the concept of "lean management" in evaluating work systems that support businesses and their associates. We will not go into detail here about this value-based model, but I encourage you to get curious and research it further yourself. Importantly, this model supports practice transformation strategies and resources, which increase professional satisfaction and caretaker wellbeing. The focus is moved to the implementation of systemic improvements in the workplace to address often-flawed processes and dysfunctional work cultures that are closely linked to caretaker burnout.

As the larger discussions develop around the substantial benefits and the need for system design and leadership to shift toward mission-driven goals in caregiving environments, how do the individuals and caretaking teams go about "nipping burnout in the bud?" Medical practitioners understand the value of being proactive with medical concerns. The same should apply to toxic burnout syndrome. It is much better to avoid the serious stages of burnout rather than to wait to hit rock-bottom.

"RED FLAGS" THAT MAY WARN THAT YOU ARE ON THE BURNOUT PATH:

- Increased sarcasm and "snark" can be early indicators of reacting poorly to things that you might historically have taken in stride.
- Physical exhaustion from start to finish of your work shift with no time where you feel rejuvenated or "in the flow" with your work.
- Detachment or feeling "bored" while you are doing your work.
- Increased procrastination or finding that it takes more effort than is normal for you to do your work tasks.
- "Mysterious" illness—stress-related health concerns that show up as GI distress, headaches, back pain, weight loss/gain, or insomnia.
- Saving up your PTO and not taking the much-needed time off for vacation or self-care.

- Guilt—you feel like you can not get all of your work done (maybe because your workload is too heavy or because you can not focus) but then you start to feel guilty about not being able to complete the work, which results in you working even more hours.
- Increased reliance upon alcohol, drugs, or food to find comfort and relieve stress. If not managed, these behaviors can lead to addictions that then need to be identified and treated with professional support.

The fundamental research done by Dr Christina Maslach et. al. in the 1970s on the new concept of "burnout" in workers led to the creation of the Maslach Burnout Scale, also called the Maslach Burnout Inventory (**MBI**).[13] It is recognized as the leading measure of burnout and validated by 35+ years of extensive research. The MBI measures burnout as defined by the World Health Organization (WHO) and it is used in 88 percent of burnout research publications.[14] There are many excellent surveys and evaluations for both individual and team wellbeing, including the MBI, that can be found throughout the website www.mindgarden.com.

Utilizing a qualitative measure to assess one's wellbeing, using such scientifically sound surveys as the MBI, allows for a "pulse point" measure in time. It also allows the individual to identify where burnout is indeed the key concern at that moment vs. empathetic distress or clinical depression.

SUPPORTIVE PRACTICES THAT MAY BE USEFUL TO CONSIDER TO SPECIFICALLY ADDRESS BURNOUT:

- Evaluate your values, priorities, and professional goals.
- Evaluate your schedule and determine whether new approaches are needed.
- Give yourself permission for an extended break (vacation … what??).
- Change roles, jobs, or work environments altogether as burnout is tied to work-related concerns, and many symptoms can be relieved with simply changing your professional circumstances.
- Increase your coping skills—learn how to reframe, practice gratitude, discover mindfulness, and possibly even meditation.
- Not all stress is bad! It is our perception of stress—change your mindset and make stress less harmful. (Eustress can actually lead to growth!)
- Neurons that wire together, fire together— creating new, healthier neuronal pathways to the hippocampus, anterior cingulate nucleus, and prefrontal cortex improves memory formation, emotional regulation, and prevents the mental decline with stress.
- Get more high-quality sleep (see Arianna Huffington's work with Thrive and The Huffington Post).
- Laugh more! Great medicine for all that ails you, including burnout!

MORAL DILEMMAS/INJURY VS. ETHICAL CONFLICT AND EXHAUSTION

To initiate this important discussion, it would be helpful to first clarify what is meant by "morals" and "ethics" in the context of this chapter. Often used interchangeably to point to what is "right" and "wrong" conduct, morals and ethics are in fact different. Morals refer to an individual's own principles and values regarding what is right and wrong. Ethics refer to rules provided by an external source, such as professional conduct or principles in religion. It is also valuable to discern that, when discussing ethics in this regard, I am pointing to what Bernard Rollins defined as the ethics that are our set of beliefs of good/bad, right/wrong, just/unjust that were informed by family, friends, teachers, community culture, and other external influences.[15] With that as a foundation, we will now explore the concepts of moral dilemmas and distress as well as ethical conflict and exhaustion.

"Moral distress" in human medicine and nursing care is defined as when one knows the ethically correct action to take but feels powerless to take that action. This can occur when a caretaker is prevented from doing what they believe to be the "right thing to do". It has been theorized that with nurses, moral distress occurs when the nurse knows what is best for the patient, but that course of action conflicts with what is best for the organization, other providers, other patients, the family, or society as a whole.[16]

In veterinary medicine, we can absolutely relate to this last statement. For example, when a veterinary caregiver is prevented from doing what they believe is the right and compassionate thing to do for a patient due to a client's financial constraints or a practice's rules around providing treatment, moral distress is likely experienced by the individual if not the caregiving team. When our values and beliefs conflict with a workplace procedure or policy (how things should be done vs. "must" be done), this can lead to a moral dilemma which can also be considered an ethical conflict. In a 2008 paper in *Nursing Ethics*, McCarthy and Deady made the point that moral distress needs to be differentiated from emotional distress, which is more generic and may occur in a stressful work environment but may not have an ethical element.[17] This concept was articulated in a concise way by Lorraine Hardingham in her paper on building a moral community: "Moral distress involves a threat to one's moral integrity. Moral integrity is the sense of wholeness and self-worth that comes from having clearly defined values that are congruent with one's actions and perceptions".[18] Importantly, when an individual experiences moral distress repeatedly, without the skills or autonomy to process or to resolve the stress that this invokes, it can lead to ethical exhaustion.

BIOETHICS

There has been greater exploration of bioethics in human medicine over the last 30 years. There is quite a bit of transferability of the knowledge and awareness gained by the many research efforts in human medicine to veterinary medicine and caregiving. As an example, the work by Dr Andrew Jameton placed a greater emphasis on ethical dilemmas than on moral distress. He stated that

In situations that engender moral distress, the ethically appropriate action is likely to have been identified. Thus, discussion of the ethical elements is less critical. Instead, addressing moral distress requires identification of social and organizational issues and questions of accountability and responsibility.

His work focused on the psychological impact of painful feelings, psychological disequilibrium, or both, resulting from barriers to performing actions consistent with one's own moral compass.[19] The key to increasing greater alignment and decreasing the likelihood of caregiver distress is the ongoing evaluation of one's own values as well as an organization exploring and clearly stating its values.

Bioethics as a field of study is still quite young for human medicine and even younger for veterinary medicine. There has already been, however, a great deal learned, and ongoing research efforts and discussions are making their way into the practice environments benefiting patients and caregivers alike. When caregivers are working in alignment with their values and integrity, there is a greater engagement in and increased quality of their patient care. The patients certainly are positively impacted but importantly so are the caregivers who feel a greater degree of compassion, satisfaction, and fulfillment in their work. This matters on so many levels for everyone involved.

A foundation of terms and concepts has now been prepared for us to move to the next level of our conversation. How do these caregiving concerns manifest in the veterinary profession and with veterinary caregivers? What are the unique concerns that exist in veterinary medicine that require attention and their own space for exploration? I invite you to take a moment to pause, reflect, and perhaps even to write some things down that occurred to you when I asked those questions.

Now, let's dive in with curiosity and compassion as we seek to understand our veterinary caregiving selves better.

NOTES

1. Skovholt, TM, and Trotter-Mathison M. (2016). *The Resilient Practitioner: Burnout and Compassion Fatigue Prevention and Self-Care Strategies for the Helping Professions,* 3rd ed. London: Routledge.
2. Newell, JM, and MacNeil GA. (2010). Professional Burnout, Vicarious Trauma, Secondary Traumatic Stress, and Compassion Fatigue. *Best Practices in Mental Health,* 6 (2), 57–68.
3. Figley, CR. (1995). Compassion Fatigue: Toward a New Understanding of the Costs of Caring. In BH Stamm (Ed.), *Secondary Traumatic Stress: Self-Care Issues for Clinicians, Researchers, and Educators* (pp. 3–28). Baltimore, MD: Sidan Press.
4. Saakvitne, KW, Pearlman, LA, and the Staff of the Traumatic Stress Institute. (1996). *Transforming the Pain: A Workbook on Vicarious Traumatization.* New York: W.W. Norton.
5. Mathieu, F. (2012). *The Compassion Fatigue Workbook.* New York: Routledge.
6. Singer, T, and Klimecki, OM. (2014). Empathy and Compassion. *Current Biology,* 24 (18), 875–8.
7. Singer T, and Klimecki, OM. (2011). Short-Term Compassion Training Increases Pro-Social Behavior in a Newly Developed Prosocial Game. *PLoS One,* 6, e17798.

8. Dowling, T. (2018). Compassion Does Not Fatigue! *The Canadian Veterinary Journal*, 59 (7): 749–50.
9. Maslach, C. (1979). Burned-out. *The Canadian Journal of Psychiatric Nursing*, 20 (6), 5–9.
10. Bauer-Wu S, and Fontaine, D. (2015). Prioritizing Clinician Wellbeing: University of Virginia's Compassion Care Initiative. *Global Advances in Health and Medicine*, 4 (5), 16–22.
11. Kraft, U. (2006). Burned Out. *Scientific American Mind*, June/July, 28–33.
12. Shanafelt T, Goh J, and Sinsky C. (2017). The Business Case for Investing in Physician Wellbeing. *JAMA Internal Medicine,* 177 (12), 1826–32.
13. Maslach C, Leiter MP, and Jackson SE. (1981). *The Maslach Burnout Inventory.* Palo Alto, CA: Consulting Psychologists Press.
14. Boudreau R, Boudreau WF, and Mauthe-Kaddoura A. (2012). *Wellbeing at the Workplace: The Need for a Psychological Counselling Service—An Exploratory Case Study.* Book of Proceedings of the 10th Conference European Academy of Occupational Health Psychology.
15. Rollin BE. (2006). *An Introduction to Veterinary Medical Ethics: Theory and Cases,* 2nd ed. Oxford: Blackwell Publishing.
16. Corley, MC, and Minick, P. (2002). Moral Distress or Moral Comfort. *Bioethics Forum,* 18 (1–2), 7–14.
17. McCarthy J, and Deady R. (2008). Moral Distress Reconsidered. *Nursing Ethics,* 15 (2): 254–62.
18. Hardingham LB. (2004). Integrity and Moral Residue: Nurses as Participants in a Moral Community. *Nursing Philosophy*, 5 (2), 127–34.
19. Jameton, A. (2017). What Moral Distress in Nursing History Could Suggest About the Future of Health Care. *AMA Journal of Ethics*, 19 (6), 617–28.

3 Being Part of a Compassionate Caregiving Community— for Better or for Worse! Examination of Our Veterinary Professional Challenges

> The capacity for compassion and empathy seems to be at the core of our ability to be wounded by the work.

> —*Beth H. Stamm*

When we identify and understand our unique challenges as veterinary professionals, we can find the opportunities, strategies, and skills that support us to thrive in our profession. Over the past two chapters, we have broadened and deepened our collective awareness of the historical and current occupational hazards of being in a caregiving profession. Having a shared vocabulary and better understanding of these wellbeing concepts for caregiving professionals overall, we are primed to more clearly discern how these same concepts currently manifest in the veterinary profession.

In my preparation for the writing of this book, as well as in my current role working alongside veterinary social workers, I have realized that there are many similarities between individuals drawn to the veterinary profession and those who choose to be human-focused caregivers. The parallel emotional and psychological impacts are entwined with being highly empathic individuals drawn to caregiving roles. Here, I will delve into those aspects in more detail and seek to translate how they apply to the veterinary realm. Additionally, there are also circumstances, ways of practicing, and caregiver concerns that are unique to veterinary medicine. Some examples of veterinary-centric issues include the impact of the human–animal bond, financial concerns, and humane euthanasia. There is also the unique dynamic for those who chose veterinary medicine because they prefer interacting and caring for animals over people but find themselves called to frequently provide care to both in vet med.

DOI: 10.1201/9780367816766-3

These are regular parts of our practice life that can have significant ramifications on personal wellbeing and on professional fulfillment.

The question at hand is: how do we become the best version of our professional selves and then peer-mentors? We, as a compassionate community, must figure out how to first care for ourselves while caring for our veterinary patients in such a way that we lay the foundation for a sustainable, fulfilling career. In turn, we can then prepare and support our colleagues to be successful in veterinary medicine as a long-term career option—professionally, personally, and emotionally. There are certainly larger sociologic and financial considerations that impact individuals, practices, universities, and organizations. All of this makes for a genuinely complex discussion. I make note of it here out of respect that I will be providing an overview of the topics here to keep the scope more general. This is not out of purposeful exclusion or out of an uninformed naivete, but to lay a common foundation for each unique individual and environment to build discussions upon.

Let's see how we can take what we know, and what we have just learned, about the veterinary profession and caregiving roles to create new approaches that support professional health and fulfillment in veterinary medicine.

ACCOUNTABILITY AND SAFETY

We need to promote a culture of caring, and this involves both physical and psychological safety for the caregivers. This is a shared responsibility between individuals and leadership. "Healthcare cultures must be a synergy of safety, quality, competence, and compassion. Too little of any ingredient will negatively impact patients and caregivers alike". This valuable conclusion comes from a recent paper that was contesting the term "compassion fatigue" with caregivers. The balance between self-care, compassionate practices to balance empathic distress, and work environments that promote these elements while also providing safety for workers promotes higher quality care for all.[1] When referring to "safety", this refers to physical safety and psychological safety.

Psychological safety for the support of the veterinary associate cannot be understated. Without this, every other aspect of the veterinary caregiving professional's career is sabotaged. This is important both in creating an environment where a colleague is compassionately supported when they are struggling in any capacity and when mistakes are made. The ability to share when a mistake is made and not fear negative repercussions or shaming is vital. This is particularly valuable to veterinary caregivers in their first few years of practice. Knowing that the practice culture is one of trust and of respect would support a collaborative approach toward learning as well as putting in safeguards to support best patient care. In this environment, novel protocols supporting best practices in workflow, patient care, and associate training are more likely to be developed and implemented.

Dr Atul Gwande's best-seller, *The Checklist Manifesto*, speaks to the need for group accountability to support a culture of increased safety for patients and for caregivers. By acknowledging the inevitability of human error, systematic protocols were created by the medical team and implemented to increase effective communication. Rather than scapegoating an individual, the approach is to work as a team to minimize the likelihood of avoidable errors. With the focus on a collective responsibility for higher quality of patient care, there is a significant decrease in medical errors and an increased sense of support and appreciation for the caregivers.

Open and supportive evaluation of when questions, concerns, or even mistakes occur is key to fostering a culture of psychological safety in environments of care and learning. Compassion fatigue and burnout are significantly decreased when associates feel valued and know that their professional growth will be supported with less likelihood of being shamed or judged. Individual accountability is encouraged when a culture of group accountability and collective learning is fostered. To better appreciate where group accountability currently exists in veterinary medicine to support safety, as well as other cultural elements promoting healthy growth and practice, it would be helpful to examine the various professional stages of development and of clinical practice.

PROFESSIONAL IDENTITY FORMATION AND ITS IMPACTS ON CLINICAL PRACTICE AND CULTURE

Closer evaluation of culture and support at the different stages of training and clinical practice is a valuable approach to clarifying where healthy practices and skills exist or have the opportunity to be developed. Breaking down the journey of professional identity development into these different stages allows for closer evaluation of where opportunities for culture improvement exist and opens the door to creating strategies for us to "do" and "be" better as caregivers—to ourselves and to one another.

I reached out to Dr Tamara Hancock at the University of Missouri who studies professional identity formation and has a special passion for and education in the science, philosophy, and ethics of curriculum development. With her assistance, I have come to understand that the contemporary classification of these development stages is:

1. Anticipatory socialization (prior to matriculation).
2. Professional student.
3. Post-DVM training (house officer/intern/resident).
4. Veterinary professional (novice < 5 years, experienced 5–25 years, senior > 25 years).

There are unique challenges and opportunities within each of these stages that we are now going to delve into a bit further.

Student/House Officer Stage of Professional Development

Anticipatory socialization is the informal training of veterinary caregivers which often begins prior to entering a formal veterinary technician program or veterinary college classroom. At whatever point an individual feels drawn to explore animal welfare and medicine as a possible career path, the journey often starts by working at or volunteering in different animal care environments. These are opportunities for exposure to ways of practicing, standards of care, team dynamics, and communication with clients.

The learning and discerning that occur in these situations are vitally important. One of my colleagues pointed out that *"The Hidden Curriculum"* is a powerful and well-documented source of cultural learning in medicine and more powerful than the written curriculum. Many find lifelong mentors in these circumstances or examples of professional decorum and patient care that stand out as the "way it should be done" in that individual's mind. There are also, however, instances where negative experiences are equally instructive. Those interactions with employers or coworkers where we felt physically or emotionally manipulated or abused, where we felt unsafe, or where we felt disrespected and undervalued by actions and words help to clarify our own values, ethics, and self-efficacy. I have my own experiences, and the words of so many of my veterinary colleagues in stories that they shared during the preparation of this book are in my heart as I write these words. Here is one of those stories that have stayed with me from my dear friend, Pam.

PERSONAL REFLECTION FROM PAM STEVENSON, CVPM (VETRESULTS):

As a lifer in the veterinary profession, I have 40+ years of experience and tales. But the whole story starts long before my working career—on November 23, 1963. When the news of Kennedy's assassination fell heavily on the world, my elementary school let us out early. Walking home with tears and a heartache that I did not fully understand, I saw my dog, Maxine, across the street. As a third grader not yet capable of predicting outcomes, I called to her, and she joyfully came running. Until the school bus crushed her. Fifty years later, that is as gripping as at that very moment. And it is what propelled me to the profession that I dearly love.

Jump forward two decades, and I am 23, recently divorced, a single mother of 2 children, one having died tragically less than a year earlier, and working as a veterinary assistant. My employer was an inspiring person, a mentor who continually pushed me to learn more, do more, and be more.

Then a registered veterinary technician (RVT) joined the pack. The RVT and I were intimidated by and disrespectful of each other from day one. Me because she was more educated. Like so many, my skills came on the job. My employer "encouraged" me to study textbooks and tested my knowledge regularly. And she because I had more experience. We circled each other with awe and fear.

One day, while passing throughthe treatment room, I noticed The RVT was about to give an apparent overdose of anesthetic. My immature, accusatory, and far from a supportive comment of "that is too big of a dose for that cat" was acknowledged and ignored. I followed that with a passive-aggressive "you're going to kill that cat", are words that, 40 years later, continue to bring me tears and heart-crushing guilt. You see—the cat died. It was MY hubris, lack of self-awareness, and total lack of conflict resolution skills that killed the cat. If, on some level, it was intentional, no amount of penance will soothe my soul.

Why do I share this terrible experience when there are hundreds of thousands of joyful ones? Well—for the same reason that I shared Maxine's demise at my hand. Because what we DO with these experiences makes us who we are. In a positive outcome, they are the catalyst to learn more about ourselves, understand and accept our faults, and make core changes to our being. These and other actions with poor outcomes formed my vision and mission to contribute to veterinary teams' personal growth, self-awareness, and essential communication skills. If even one precious soul is saved from being sacrificed on hubris, fear, and lack of courage to do the right thing, always, my career has been a success.

If our selective perception of the positive experiences outweighs the negative, and the draw toward veterinary medicine remains strong, the determination to get into a formal training program persists. The path to successful application and acceptance to veterinary school requires genuine resilience, passion, and focus. Competition is fierce to get a seat in a veterinary classroom. As a result, those that secure a prized spot trust that the attributes of high-achieving individuals (e.g., striving, determination, competitive mindset) that served them up until that point will be the same needed to successfully "survive" veterinary school. Most veterinary students have high expectations of themselves, of the training environment, of their instructors/mentors, and of the material that will be provided to them to best prepare for clinical practice. Perfectionism in its maladaptive form is often associated with high-achieving individuals. It is an important bias that impacts the wellbeing and learning of the individual but also can be imposed in a judging way upon others. This is an important concept that we will visit later in the chapter as to how perfectionism manifests and how it can negatively impact learning, professional and personal development, and mental health.

The professional student stage is where individuals are likely to be the most receptive to new ideas and ways of being. The mindset is one of genuine curiosity and there is increased receptivity to hearing guidance and wisdom from mentors and instructors. The sense of being accepted into this new "club", where all of the invested time and energy will now pay off, is exciting. There is an unspoken trust that the students will receive what they need to at least have the fundamental building blocks for what will be a successful career in veterinary medicine. This is where knowledge and skills around wellbeing can complement clinical training to prepare us to be successful, more fulfilled clinicians.

I have asked several colleagues who are currently involved both in development of veterinary curriculum and in the teaching about the receptivity of students to nonclinical lectures and discussion forums. Many students do not yet have the experiential awareness to understand why these topics are as important as anatomy and histopathology in preparing them to be successful practitioners. The fact that these are not the courses that make the grades that keep them in school may have led to a historical reluctance to engage with this material with the same level of interest given to core clinical courses. In more recent student assessments, surveys, and conversations, however, there is a "growing consensus of students who perceive the need and demand structural or curricular change to address concerns about wellbeing and other humanistic essentials" (personal communication, Dr Tamara Hancock).

The onus is on the training community, and us as peer-mentors, to demonstrate the practical value of incorporating soft skills and wellbeing knowledge as much as providing clinical content to these eager students. There is the potential for meaningful synergy between the academic professors and the peer-mentors from the clinical environments to share *why* all that is being taught matters! With the focus on a progressive, strategic approach to the preparation of our veterinary caregivers of today, curriculum content and the means of teaching needs to be investigated and, in some instances, updated.

Historically, the formal veterinary curriculum was intentionally broad in an effort to prepare students for a variety of possible career paths within the veterinary profession. There were at least seven "core" species and their unique variances that may impact care and treatment decisions. For those students likely to pursue careers in pathology, zoo/exotic medicine, or epidemiology, additional coursework and experience would be expected. This approach was appropriate for the time when most veterinarians might practice some form of mixed animal medicine upon graduation, and it remains true for those who serve more rural areas. However, the significant increase in medical knowledge for each species as well as the advent of specialty medicine in private practice in the 1990s has contributed to the need to consider whether that approach is simply too broad and impractical.

The universities, the accreditation forums, the American Association of Veterinary Medical Colleges (AAVMC), and national organizations are currently evaluating where and how to make meaningful changes. In addition to considering the curriculum content, the format of information delivery is also being considered. There will likely remain certain subjects that are appropriate to teach in the historical pedagogical method of having a live instructor delivering content to the entire class. Some of the younger veterinary schools are experimenting with how early the practical application in clinical environments is incorporated into the training time. Smaller student cohorts supporting group discussions and learning are being increasingly utilized. Most schools are also offering an increased number of electives or complimentary degrees alongside the veterinary degree, e.g., leadership, business acumen, holistic care, education. Although change to such long-standing approaches to veterinary curriculum may be slow, it is definitely developing and will positively impact the holistic preparation of future veterinary professionals in both clinical acumen and sustainable caregiving practices.

PERSPECTIVE FROM DR MICHAEL SCHAER, LONG-TERM CLINICAL PROFESSOR AT THE UNIVERSITY OF FLORIDA:

Are we continuing to teach our students how to reason or are clinicians jumping to the finish line without the slow and methodical principles involved with objective learning? Are we teaching the art of medicine or have we wrongly adhered to nothing but evidence-based medicine which sometimes causes colleagues to blind themselves in such a way that they cannot see the forest through the trees? I have seen both ways of teaching, and the potholes created when one is emphasized over the other. Let's not forget "Uncle Mikey's Maxims", as these will get them through many of the dilemmas of real-world general practice. (See Appendix).

In addition to these valuable considerations, research-driven curriculum changes supporting humanistic factors and preparation for professional practice are being developed to compliment the clinical content. In speaking with Dr Tamara Hancock at the University of Missouri, she shared that there is an entire science and methodology to developing curricula. In addition to updating of content and teaching approaches, consideration must be given to student outcomes on a broader and more practical range of criteria. Namely, in addition to the clinical knowledge that the student needs to start ethically practicing quality medicine, preparation to navigate the innate complexities and challenges of clinical practice, working with clients, and self-care should be provided. Dr Anne Quain at the University of Sydney also pointed out to me that the voice of the student in decision-making by education institutions has shifted, particularly as tuition fees have significantly increased. Student feedback on their professional interests is strongly considered in the development of new curriculum topics.

At Bristol Veterinary School, Dr Lucy Squire and several of her associates now lead seminars and small group activities in all five years of the BVSc course. The tools and techniques are intended to promote mental health and wellbeing. In a recent article published by Dr Squire and her team, the "Mental Wellbeing Toolbox" that they developed and draw from to guide their teaching and discussions was publicly shared.[2] The handbook that they created to share more broadly includes details of all topics covered in the Mental Wellbeing Toolbox coursework, including a complete list of references from the literature review.

(Note: The Mental Wellbeing Toolbox Handbook is freely available online at www.bris.ac.uk/vetscience/media/docs/mental_wellbeing.pdf).

Veterinary Classroom Culture

If we are serious about promoting a culture shift in the practice environments, it starts in the classroom. The application process and admissions criteria even before matriculation may apply selective pressure toward particular personal or relational attributes or tendencies. As an example, perfectionist tendencies can contribute to healthy work ethics and organizational skills but, when out of balance with a strong sense of self-worth, may interfere with growth and connection to community. We will spend more time defining and exploring the impacts of our perfectionist tendencies in the veterinary profession later in this chapter.

Conversations and initiatives are still developing around the broader topics of equity, inclusion, and diversity. Our veterinary profession has struggled with diversity issues. This is another important contributor to the culture shift in our profession toward supporting psychological safety for all veterinary professionals. Historically, veterinarians were white males with a mix of males and females in the parastaff roles. The male:female ratio finally hit 50/50 in the late 1980s. This was clearly an important tipping point, as the current veterinary classroom is approximately 87 percent female. Over the last 20 years, individuals from a diverse array of socioeconomic and academic backgrounds, historically underrepresented groups, and social identities have been increasingly recruited to apply to veterinary programs. However, according to the 2019 Bureau of Labor Statistics report "Labor Force Characteristics by Race and Ethnicity", 92.8 percent of veterinarians in the workforce were white, making it one of the least diverse healthcare professions.[3] This demonstrates the enormous opportunity to develop and foster environments of belonging in our profession.

> Initial efforts to support equity, inclusion, and diversity in all of its manifestations have fortunately led to the increase in formation of student chapters supporting LGBTQ and different cultural heritages over the last 15 years. As of the writing of this book, examination of the websites for the North American, Australian, and British veterinary colleges reveals there is currently representation of forums for Equity, Inclusupport to bolster feelings of safety and of inclusion at almost every school. Student chapters for VOICE (Veterinarians as One Inclusive Community for Empowerment), Broad Spectrum Veterinary Student Association, PRIDE VMC, the Australian Rainbow Vets and Allies, as well as many others supporting wide ethnic backgrounds are the first of many necessary efforts to encourage the veterinary professional community of the future. Please check out Chapter 6 for a more complete international list of EI & D platforms that support veterinary students, faculty, and practicing professionals.

Novice Professional Phase and Professional Development

This is where the rubber meets the road because academic knowledge is now applied to real-world scenarios, and they are not what we expected while in the classroom or during clinical rotations/externships. All too often, students are taught the "gold standard" or "best practice" without the context of the practical and financial challenges that will inevitably be faced in clinical practice, e.g., the family that does not have the financial means to pursue recommended diagnostics and treatments. These new practitioners are experiencing, for the first time, having to be responsible for a patient's life, for building trust with clients, and with working effectively and compatibly with a team of other caregivers.

As we identified earlier, the stage of the novice professional refers to the first 1–5 years of clinical practice after completing the formal training program. In veterinary medicine, this would apply to both our recent veterinary technicians and clinician graduates. This stage of professional development is critical in the creation of professional identity and contributes largely to the trajectory of an individual's personal and professional wellbeing. Dr Liz Armitage-Chan and others have researched and written about the formation of professional identities and how these can influence a graduate's mental health as well as sense of job satisfaction.[4]

Although better understood in other caregiving professions, this stage of professional development is a learning opportunity for our veterinary community. Additional research done specifically on novice practitioners in veterinary medicine confirms that there is increased psychological distress and depression in more recent veterinary graduates.[5]

Many young veterinarians find themselves struggling with the application of academic knowledge to the "real world" and coping with the broad range of challenges innate to "the novice" in clinical practice. There can be significant anxiety around not feeling prepared for all the nuances of clinical practice and finding that compassionate support and mentorship is often promised but under-delivered. This can leave the young practitioner to experience debilitating thoughts that they were somehow "not enough": "I am not smart enough. I do not know enough and never will. I don't know how to communicate this the 'right' way!" These young colleagues may even question at this point whether they made a mistake by choosing this particular professional path.

If significant school debt is also a consideration, the young graduate may experience a sense of being trapped with no clear way out. Concurrent thoughts of having knowledge and skills that are unique to this particular job/role that would not easily transfer to another job/role (even if within the veterinary profession) may contribute to both feeling stuck and exhausted by the anticipated energy it might take to start anew. Feelings of hopelessness and isolation may then arise. If this individual is in a practice environment that, like many practices, has a culture of perfectionism and judgment rather than of supportive community, the development of compassion fatigue and of burnout becomes more likely.

STORY FROM DR MEREDITH KENNEDY (30+ YEARS OF ER MEDICINE): "THE DYSPNEIC CAT"

My patient was turning blue, and my blood pressure was through the roof, having only a matter of months under my belt since graduation. It was a ten-year-old cat with labored, open-mouthed breathing, purple mucous membranes and loud, stridorous rasping with each breath. The worst part was the look of frantic agony on the kitty's face while he sat perfectly still, unable to do anything else but labor to breathe. I guessed he was having an asthma attack, but he wasn't stable enough for radiographs, so I gave him terbutaline, steroids, and oxygen. He was still asphyxiating, and with a sinking heart I realized he was going to need a tracheostomy, or he would die. He had arrived in a state of laryngeal edema, and steroids weren't going to be enough. I placed an IV catheter, easily done since the cat sat like a statue with no one needed to hold.

However, my boss, a board-certified surgeon, would not take kindly to me doing any surgical procedures without him, so I called him. He ordered me to wait for him to arrive, do not touch the cat until he's there. Period. I transferred the cat to an oxygen chamber and waited anxiously. The cat continued gasping and turning purple.

Once in the building my boss scowled at me and looked at the cat in oxygen. "Where are the X-rays?" he demanded. I explained that I hadn't gotten any yet since the cat was too unstable to place lateral. "If he's stable enough for an IV catheter, he's stable enough for X-rays", came the answer. It was no use trying to explain that I had placed the IV catheter in 60 seconds, with zero stress to the cat.

Predictably, the radiographs were stressful, but the cat survived. "Asthma", grunted my boss, glaring at me. Anyone could see the cat had asthma. He ordered the cat kept in oxygen and treated medically. I tried to talk to him about the upper airway obstruction, but he wasn't having it. The X-rays only showed the chest and didn't include the larynx. The cat does not need a tracheostomy. Period. He went on with his day, and I finished my shift. I couldn't stop hovering over the poor cat, miserably gasping in the oxygen chamber.

The overnight emergency doctor, older and much more experienced than I, started her shift and we did our rounds. Stopping in front of the oxygen chamber she watched the dyspneic cat and asked, "Why doesn't he have a tracheostomy?"

Boss says it's asthma. He doesn't need a tracheostomy. I was about ready to cry, I could see that the cat was dying. She rolled her eyes. She was an outstanding doctor, fast and competent, and she of all people could intimidate the boss man. "We'll see".

I dragged myself home and lay awake tossing and turning, haunted by the look of desperation on that cat's face. I couldn't sleep. The phone rang and it was the overnight doctor. "I thought you'd like to know I just placed a tracheostomy in your cat. He's doing fine".

What? How did you manage that? The boss said no way! "I just called him and told him the cat was dying, and I'm placing the trach tube".

The cat was indeed fine. We left the trach tube in place for 48 hours while the laryngeal edema subsided, and the cat ate, drank, groomed himself, purred, and solicited head noogies. The boss man's only comment was, "Seventy-five percent chance the cat's going to die anyway".

The tracheostomy tube was removed at 48 hours and the cat went home on oral medications for managing his asthma. He was fine. I was not. I could not wrap my mind around the fact that my boss had been willing to let the cat suffocate to death, rather than help me learn how to treat him properly. As a young recent graduate, I had not had the experience nor the chutzpah to handle this emergency. I could have used a mentor, but what I got instead was a bully.

The good news is that there has been a worldwide increase in awareness of the need for proactive preparation and for in-practice support for novice practitioners. Veterinary colleges have increasingly added wellbeing conversations into the curriculum as well as incorporating discussions around topics such as ethical conflict, challenging conversations, conflict management, and self-awareness/accountability.

In formal house officer training programs and in private practice environments, consistent and compassionate peer-mentorship has also been evolving with support for the mentors developing as well.

It is key that practitioners of all stages of development experience a culture of support around the "experiential humiliation" that occurs from not having enough experience yet to have answers right away, to be as efficient as they wish to be, or to be as skillful in their communications with clients and colleagues as they will be with some years of life experience under their belts. There is learning that occurs from each case, from the veterinary patients themselves, and from each other for the entire duration of our clinical career.

It helps to embrace that "failure" is simply another way of saying that the actions that we took did not yield the outcome that we desired or that we determined qualified as "success". Rather than experiencing the self-sabotaging feelings of shame, doubt, guilt, or disappointment, the opportunity is present to reframe the experience as simply an unanticipated growth and learning opportunity. Nelson Mandela stated wisely that "I never lose. I either win or I learn". It is in these moments that we also develop the capacity to be humbly reflective, which bolsters both future self-confidence and resiliency.

SPECIFIC VETERINARY WORKPLACE CONCERNS FOR VETERINARY PROFESSIONALS

One of the things that draws individuals to the path of veterinary medicine is that there are such a wide variety of applications of our veterinary education and degrees. There are so many ways of supporting animal health and welfare as well as contributing positively to human health throughout the world. There are work-related issues that impact a majority of veterinary caregivers regardless of the veterinary segment being considered, e.g., companion animal vs. food animal medicine. Each veterinary environment has uniquely positive attributes as well as stressors. In preparation for this book, I intentionally sought out opportunities to speak with veterinary colleagues that work in different countries and in a variety of work environments. These conversations, as well as my review of veterinary literature on wellbeing issues in the profession, helped to clarify which work-related issues are more ubiquitous and which concerns are unique to particular workspaces and communities.

The ecosystems themselves where veterinary professionals practice have an enormous impact on the degree to which those associates thrive or suffer throughout their veterinary career. When fortified with awareness of these conditions tied to our work environments, we are better informed and capable of identifying positive solutions. Individually and collectively, we can then be more effective in bringing about impactful change where and when it is needed.

This chapter will largely be dedicated to all the ways that we are all much more alike than different throughout our varied veterinary communities. First, however, I want to point to some concerns that may be considered unique to certain veterinary environments—not exclusive to them, but perhaps more frequently experienced in that veterinary segment. The following "callouts" are from conversations, publications, and from my personal experience. I admit that the statements and comments are generalized and subjective, as they are perceptions. However, you have heard this statement likely a thousand times: "Someone's perception is their reality". With that, I am willing to bet you might resonate with some of these from your own experience!

Small Animal/Companion Animal Medicine (GP and ER)

- Small animal practice culture—a sense of profit above "people-first" culture is perceived in larger practices (corporate more so than small, privately owned businesses).
- Academic teaching hospitals—martyrdom and stoicism as coping mechanisms are a historical "part of the job description" being taught to veterinary students and house officers. The "I did it and survived, so you can, too!" or "Suck it up, buttercup!" mentality.
- Shelter medicine—the perception that the wellbeing of the animals receives far more consideration and compassion than the human caregivers working to help them.
- Unclear job descriptions are common (important boundaries and clear expectations are not specified).
- Incivility between veterinary colleagues at work, the infamous "us vs. them" phenomenon (e.g., specialty vs. ER, front desk staff vs. treatment staff, night staff vs. day staff).
- Bullying in the workplace. Occurs between colleagues just as much if not more than what is experienced from clients bullying the veterinary team. Note: A New Zealand study revealed 16.2 percent of veterinarians had been bullied at work and found that this was more common in female than in male respondents.[6]
- Physical space is at a premium. The layout of the practices historically does not support opportunity for quiet contemplation, reading, creative thinking, assimilation of knowledge and experiences, or specifically to recharge. This results in a sense of not being invested in as humans with human needs and contributes to disengagement and burnout.
- Workplaces described as the "bullpen"—loud, chaotic work spaces.
- Clinicians seen as "cash cows"—the primary drivers of revenue for the practice which then pays for the salaries of others (including the practice owners and leadership). You will hear this from many types of small animal environments but particularly emergency care.
- Staffing—progressive decrease in available nursing staff and clinicians combined with the recurring cycle of training newly hired staff members is demoralizing and exhausting—there are not enough appropriately trained/prepared staff per shift. When the team is running lean, feelings of guilt around taking any time off increases (even while working your shift). Enormous impact and ethical conflict that these circumstances engender

have contributed to a decrease in the quality of medical care for the patients. These issues are significant contributors to compassion fatigue, burnout, and physical exhaustion experienced by the entire team.

- When chronically understaffed, allowing dysfunctional work environments to persist also does not encourage associates to stay.

LARGE ANIMAL MEDICINE

- Ambulatory only vs. ambulatory + practice environment (mixed animal practitioners)—Can provide a sense of autonomy and freedom and allows for more time outdoors.
- The time spent driving to/from appointments can be physically taxing.
- Potential for significant moral distress if the patient has already died by the time the clinician arrives on site.
- Flexibility of schedule and of client response for some vs. "always on call" for others (no boundaries at all around their personal lives).
- Physically demanding for almost all types of large animal medicine. Many of these veterinary patients are large and heighten risk of physical injury to the veterinary caregivers.
- Most of these positions do not have high income potential +/- benefits for associates (unless practice ownership is considered).
- Working in isolation (decreased community support, discussion of cases/ patient care) can contribute to an increase in mental health challenges and in burnout (especially for women who wish to have families).[7]
- These factors, the hard work, and the remoteness of most of the jobs contribute to the challenges of recruiting veterinarians to these positions.

POULTRY MEDICINE

- Roles vary quite a bit based on the employer (e.g., USDA vs. food company).
- Opportunity to be integral in the creation and implementation of protocols that directly impact animal welfare as well as food safety.
- Job descriptions can vary widely and require clear communication of expectations from both veterinary associate and employer.
- Working with larger companies provides opportunities for positive work–life integration, healthy income growth, and professional development.
- Challenge of having to euthanize a large number of animals if there is evidence of contagious infection (e.g., avian influenza)—ethical dilemmas associated with mass slaughtering, or culling of animals, even if for the protection of other animals and/or human health.
- Ethical conflict can occur if protocols of inspection and understanding of clinical symptoms in animals or physical evidence of infection in animal products are not clear—may result in a large amount of meat being wasted due to being determined "not wholesome" or "not safe for consumption".

LAB ANIMAL MEDICINE

- Opportunities to develop other professional skills, such as being an effective leader, managing a team through a research project, and navigating uncomfortable conversations between parties with differing agendas.
- Stress around having too few individuals working on a project with rigorous time and work expectations.
- Work–life integration is possible with thoughtful hiring, staffing, and boundary creation ("on call" actually exists in lab medicine as well, e.g., trauma to the animals in the lab or a rodent dystocia).
- Opportunity to work alongside or in conjunction with other teams on important research endeavors (e.g., CDC, NIH, universities).
  Note: Dr Larry Carbone's excellent book, *What Animals Want* (Oxford University Press, 2004), addresses tensions between animal advocates and researchers—with lab animal vets often caught in the middle.

ZOO AND CONSERVATION MEDICINE

- Highly politicized environment: decisions around animal care can be heavily influenced by nonmedical team/caregivers (e.g., the board of directors) which can make quality of life discussions particularly fraught.
- Animals and their wellbeing/welfare are visible to the public.
- Developing and maintaining trust between the veterinary caregivers and the zookeepers (to facilitate communication and increase treatment compliance) is vital and requires vigilant attention.
- Potential for contentious conversations and ethical conflict around how funding is allocated in the running and upkeep of zoo and wildlife parks—public infrastructure vs. animal care vs. environmental upgrades/enrichment vs. charismatic species receiving more funding than "less sexy" species (e.g., conservation efforts fpandas versus dung beetles, which are currently one of the most threatened terrestrial animal species!).
- Potential for moral stressors and ethical dilemmas, e.g., individual animal vs. population welfare, as well as deep gratification/satisfaction around conservation medicine and education opportunities.

ACADEMIA

Opportunities to teach/mentor and to work in an environment of innovation can be invigorating for some and exhausting for others. One must enjoy teaching and be prepared for the scope of administrative responsibilities.

- Full-time faculty have to also balance the ongoing demands of creating/completing research endeavors in addition to teaching and clinical obligations.
- Slow decision-making due to "red tape" in academic environments and/or inefficiency around patient care in a teaching hospital are recurrent reasons

of why individuals determine to pursue private practice or self-employment opportunities.

- Schedule of on-clinic, off-clinic, and simply "off" (as during student vacations) works well for some and not well at all for others (Note: There may be ongoing administrative and planning duties that are expected to occur during the university holiday times).

CONCERNS FOR ALL VETERINARY PROFESSION SEGMENTS

PROFESSIONAL COLLEAGUE RELATIONSHIPS/COMMUNICATIONS

Historical culture of judgment (and shaming…)—low psychologic safety for workers:

- Fear of making and having to explain medical mistakes or "near misses".
- Fear of being perceived as "stupid", "less than", "not capable" if questions are posed to colleagues.
- If supervisor/mentor is overly critical without providing compassionate support, the student or younger clinician will be slower to, or less likely, to expand knowledge/skills outside of their "comfort" zone of habitual/known approaches. This results in stifled growth and confidence.
- The WSAVA (World Small Animal Veterinary Association) & FECAVA (Federation of European Compassion Animal Veterinary Associations) published *"Global Principles of Veterinary Collegiality"* in January 2021 as a call for professional respect and decorum to be demonstrated inside the schools and in practice environments toward professional colleagues. (www.wsava.org).

Schedule Issues

- Schedule varies by role (novice vs. senior professional, house officers, GP vs. ER vs. specialty), segment of veterinary medicine, work environment, and number of years in the profession.
- Arranging and taking of paid-time-off (PTO) when working with smaller group of associates, on your own, or in largely rural areas can feel impossible (and ethically challenging with thoughts around "If I am not here to care for the patients/clients' needs, then who will?").
- Important conversation around boundaries (between personal lives and professional expectations from themselves and from employers):
  - Long working hours
  - On-call responsibilities
  - Administrative duties occupying large portions of the veterinary professional's time when working
  - Inflexible schedules and/or requiring that someone find coverage for their shifts if going on vacation, if they are ill, or if there is a family emergency (sick child or elder).
- Schedule concerns are a significant contributor to burnout for veterinary technicians and clinicians. Long days with minimal breaks due to high caseload and patient treatments are physically, psychologically, emotionally demanding. Additional time invested in medical record writing, rDVM communications, client communications can lead to being in the hospital hours after scheduled shift time which leads to less recovery time between shifts and increased resentment of personal time infringement.
- When time off actually occurs, many colleagues share that they are so exhausted that the time is spent merely recuperating; that there is no energy available to do more than the necessary domestic duties +/- care of family.
- Disregard for family responsibilities with the perception of little importance given to the vagaries and unpredictability of childcare needs.

CLIENT RELATIONSHIPS

- Unrealistic client expectations of veterinary staff and of their veterinarians (availability for patient care, cost of care, value of services vs. acceptance of financial commitment)
  - Client expectations entangled with the human–animal bond: humanizing the relationship with the pet results in increased expectations of caregiving, but often the client is unrealistic or uninformed about the expense involved with this care.
  - Client expectations vs. veterinary caregivers who try to fulfill the false narrative of being "all to everyone", to "fix everything", to "not let anyone down", to be smart and capable, and to be *valued* for our role in caring for their pet. (I love my friend Dr Anne Quinn's phrase *"the dangerous myth of omnicompetence"*—dangerous to vets and to our clients!).
  - Feeling valued by the clients is a *BIG* deal! We attach so much of our worth as a human on this planet onto our capacity to successfully care for the veterinary patients and to receive approval by our professional peers and by the animal guardian(s).

Now, get me a diagnosis!

How I feel sometimes working within a client's constraints

- Impact of technology on the veterinarian–client relationship.
- Appropriate boundaries are pushed because of client expectations in the current "culture of immediacy"—expecting e-mail and/or text responses from their veterinary practice/veterinarian when they have concerns or

questions: "Is this normal?", "Should I be worried?", "What would you do if this were your dog/cat/horse?" (Note: This is happening right now, in 2020–2021, with COVID protocols, resulting in novel approaches to client communication).

- Fear of client complaints—can drive responses to inappropriate requests from clients and efforts to placate highly emotional (and vocal) clients. Social media has exacerbated this issue exponentially in recent years. Can also lead to "defensive medicine" and a large amount of time invested in medical record-writing to CYA (cover your a…).
- Bullying behavior or emotional blackmail, e.g., "If you really cared about animals, you would…". Using our compassion and concern for an animal's welfare against us as kind, accommodating professionals.
- Cost of veterinary care—pet insurance awareness and purchase has improved but continues to be underutilized by animal owners. The cost of pet insurance, however, can make it unavailable to some owners. "Wellness plan" options with general practitioners have improved preventative care and diagnostics. Emergency and Specialty Care—limited options to finance the care for these pets adds to the client's stress and can significantly impact the treatment plan for the patient (economic euthanasia vs. humane euthanasia).

Interestingly, in both the 2018 and the 2019 Merck studies exploring veterinary wellbeing, survey results revealed that large animal practitioners (particularly those practicing mixed animal medicine) are generally less stressed by their work and have better work–life integration when compared to small animal/companion practices. These practitioners' responses pointed to fewer financial concerns, a greater sense that their work–personal time is balanced, and generally less incidence of suicidal ideation/suicide in this population of veterinarians.[8]

EXPLORING THE IMPACT OF TECHNOLOGY ON THE VETERINARY PROFESSION

Along with medical advancements, progressive applications of technology allow for increased efficiency and quality of care for our veterinary patients. Advances in other forms of technology now enable communication with clients and others in the profession with speed and convenience. Gone are the days of hard copy medical records, fax machines, and perhaps the days of printed lab results from diagnostic equipment. Even our imaging capabilities have shifted to primarily digital platforms rather than requiring x-ray film processing. The benefits of diagnostic, medical, and surgical technology that allow us to provide a higher level of care to our patients with more accuracy, safety, and efficacy cannot be understated. The technologic advances that are a bit more complex are those associated with our medical record systems and telecommunications, including social media platforms.

Let's explore how these last two technological advances have augmented current clinical practice and, at the same time, have negatively affected veterinary caregivers and practice culture.

PATIENT INFORMATION MANAGEMENT SYSTEMS AND MEDICAL RECORDS

I remember well the days of hard copy medical records before, during, and after my veterinary training. The physical records could be misplaced, damaged, and the handwritten content could be hard to translate (unfortunately the reputation of most doctor's handwriting being illegible is true!). All of these concerns as well as having to "Xerox" the medical records and treatment sheets to share with clients and other veterinarians were wrought with potential for compromised patient care and inefficiency. The advent of the computerized patient information management system (PIMS) was initially met with trepidation and resistance by the veterinary community. Via trial and error, more veterinary-centric and appropriate systems were created over the course of a decade or so. Over time, these systems have brought some genuine improvements in patient care, particularly through fewer human errors in digital medical records and treatment sheets (no handwriting to decipher!). The ability to share patient information more readily with clients and between professional colleagues/teams also supported the best interest of the patient and is far easier than when dealing with physical records. The digital medical records also greatly facilitate sharing of records for referral of cases and processing of insurance claims.

As veterinary medicine, diagnostic equipment, and technological understanding has advanced so have the quality of the patient information management systems. Further innovation even in the last five years has resulted in fewer manual entering processes, increased charge capture, and a marked increase in efficiencies in creation/communication of treatment plans with the entire care team. These platforms have also been developed to allow all teammates to have visibility into real-time patient status to improve client and referring veterinarian updates from all staff members.

The best advancements in these patient information management systems have come from veterinary professionals who have had direct experience with quite a few of the varied technological platforms available to veterinary practices. Each of them had many years of clinical experience that informed their respective decisions to tackle recurring and persistent problems in the veterinary environments that were impeding best patient care, client communication, and veterinary associates' wellbeing. Here are a few examples:

1. Dr Ivan "Zak" Zakharenkov created **SmartFlow** in 2016. This was the first workflow optimization system that allowed many processes to be digitized, mobile, streamlined, and more accurate. He was seeking to decrease inefficiencies around treatment sheet creation and updating, decrease medical mistakes due to misreading or mistranscription, and increase ease of communication between workmates as well as with clients. Digitized consent forms, patient charts, dental charts, and anesthesia monitoring sheets support the move toward paperless practice and embedded calculators to reduce medical errors and save time for the staff.

2. Dr Caleb Frankel created **Instinct** in 2017. He qualifies himself as an "accidental entrepreneur". I would say that is super humble; as an ER veterinarian with many years of clinical experience, he recognized the impact of inefficiencies on the entire team's wellbeing and created Instinct. The goal certainly was to provide sustainable practices of accurate communication among teammates in the comanagement of patient care and to support treatment record creation and replication with fewer errors which would promote patient safety. Instinct has since further grown as a platform, connecting the team with more ease, robustly supporting patient safety, and now supports learning and professional development for all levels of the veterinary care team.

3. Dr Mark Olcott created **VitusVet** in 2014. This is a first-of-its-kind veterinary practice management software system intended to provide efficiency, increase productivity, and streamline communication with clients and between veterinary teams. Dr Olcott and his team hope to bring ease and efficiency of communication regarding medical records and care updates between the pet owner and their primary care veterinarian. The resources that they offer continue to grow as client and veterinary practice needs evolve. They are currently facilitating touch-free payments and some really cool "remind/pay/deliver" software to help practices effectively compete in the marketplace and save their pharmacy business.

SmartFlow and Instinct both seek to address the exhausting and unnecessarily inordinate amount of time that medical record creation/updates were taking the veterinary techs and clinicians. Speaking from experience, I have used no less than 8 different PMSs in my career. There were significant improvements with each system I used between 1996 and 2015, but there was a ridiculous amount of redundancy, inefficiency (too many clicks or having to go in/out of different screens), manual input (e.g., lab data), and potential for human error. The amount of time spent on medical record and treatment sheets has been, and currently still is, a huge complaint for veterinarians. When there is more data entry than medicine being practiced, it wears on the clinicians. There is increased frustration and distress resulting in a decrease in morale. The stress experienced by staff whenever practice leadership determines to change over to a new PMS or to add a new platform also needs to be considered. As a professional community, we are still learning how best to train teams efficiently, with the fewest obstacles, to maximize use of the new platforms in support of patient care, while making sure everyone feels the value and positive impact on their work satisfaction and improved quality of care being provided to the patients.

The good news! We are blessed to have smart, capable, determined individuals that are creating systems that embrace patient care, patient safety, learning for novice professionals, efficiency for all involved in patient care, and ability to easily communicate with clients and other veterinary practices. Genuinely, everyone wins!

These newer platforms also support charge capture and invoicing which contributes even further to support of practice profitability. The stress on the veterinary team and time commitment of implementation of new systems, however, must be considered when choosing what, how, and when to bring these new systems into the practice. When approached with respect, appropriate support, and clear communication, the pros often outweigh any downsides. All of this proves that when quality medicine is practiced with the right tools and a great clinical team, the business and the people can thrive.

WORDS OF WISDOM REGARDING TECHNOLOGICAL ADVANCES IN VETERINARY PRACTICE FROM MY COLLEAGUE, DR JULIA JONES:

Technology has been a Faustian bargain. We have access to more information at our fingertips than we ever imagined in 1985, but it comes at a cost. All this technology is a time suck of phones, computers and a constant contact with extraneous advertising, sketchy information, and messages that we have to keep clicking only to realize we need to delete it. We have telehealth, telemedicine, and Google with clients getting their information from different sources. We are in danger of losing the personal touch. The veterinary-client-patient relationship is the basis of our profession and sets us apart from the others. It is what makes us relevant. Manage your time, but keep hold of that personal interest in your clients and patients that keep them rebounding to you. Call them back. Guide their choices. A grateful client is a medicine for our psyche. It may not make the stresses and the "bone pile" go away, but it puts them in perspective.

At this point, I want to dig into digital technology and the interaction with patient owners. There are, again, absolutely upsides and downsides to what advances in these technologies have brought to the veterinary profession. Some of the upsides of texting and social media: excellent for promoting practice brand; increased ease of communication with clients (e.g., texting appointment reminders and patient status updates); networking among professional peers; telehealth consultations (enormous increase in utilization during 2020!). The downsides are, unfortunately, numerous and significantly impactful to individual associate wellbeing and reputation, to practice reputation and success, and to the perception of veterinarians and veterinary medicine by the general population.

IMPACT OF SOCIAL MEDIA

Cyberbullying, venting, shaming in a public forum, and removal of appropriate professional boundaries are the issues that are of most concern. These topics have been addressed in a variety of forums—articles, blogs, podcasts, lectures, in veterinary textbooks—and each has shared valuable words of wisdom and sought to increase awareness and responsibility. I have shared some of these in the "recommended readings". I found the section on the impact of social media on the veterinary team in Drs Siobhan Mullan and Anne Fawcett's (Quain) recent book on veterinary ethics provided an excellent overview of points for consideration. Inappropriate, unprofessional, and/or thoughtless use of social media can have widespread impacts on "professional reputation, collegial relations, and career opportunities".[9]

Professional organizations and practices have created forums where open, clarifying conversations around repercussions of social media use/comments have been created to help safeguard against negative impacts on individuals and teams. As an example, Dr Quain shared that the 2010 Australian Medical Association published *Social Media and the Medical Profession: A Guide to Online Professionalism for Medical Practitioners and Medical Students*. Issues that are addressed include breach of client confidentiality, slander, blurring of personal and professional boundaries, and impact on future employment opportunities. The American Veterinary Medical Association (AVMA) "Social Media Community Guidelines" supports similar tenets and includes a segment on "No Playing Vet" stating:

> We are unable to provide health advice for your pet over the internet—it is unethical and often not legal for us to do so. An online resource is not a substitute for veterinary examination and care, and online interaction is not a valid Veterinarian-Client-Patient Relationship (VCPR). On occasion, we post (or allow to be posted) questions from veterinarians asking for input from our Facebook audience. The veterinarian who made the request is responsible for his/her own diagnostic/treatment decisions.[10]

Whether intentionally or inadvertently, use of social media can have significant and distressing impacts on individuals. Clients may lose trust in their veterinary provider, relationships between peers or between clients and providers may be damaged, lawsuits can result, and an individual may be traumatized if they are personally attacked/disparaged in a public forum. The lines between private and public can be blurred, making it challenging for professionals to navigate this space. Emotions inevitably run high when someone is disgruntled, angry, or feels that they have to speak up on a concern that they view as ethically distressing (e.g., perceiving that inappropriate patient care was provided by an individual or a practice).

The shaming, vitriol, and damaging comments that can result from virtual bullying can be truly devastating. Cyberbullying has been a contributor to individuals reporting mental health concerns or suicidal ideation. It is considered a form of psychological violence that can cause serious damage to both organizations and to individuals. The AVMA reports that approximately 1 in 5 veterinarians surveyed in 2014 reported that they had themselves or knew colleagues that had been victims of incidents that ranged from posting negative reviews to threats of physical or financial harm. Almost one-half (48 percent) of veterinarians who had been victims of cyberbullying had considered a career change because of the incident, and there are documented cases where veterinarians have been bullied to suicide.[11] Michelle Winter also wrote an excellent article in the Spring 2020 issue of *Today's Veterinary Nurse* that details how to recognize and navigate cyberbullying in both one's personal life and in the veterinary workplace.[12] Eric Garcia has been a valuable resource as a speaker, consultant, and advocate for veterinary professional wellbeing, particularly as it pertains to understanding the positives and "dark sides" of technology in veterinary practices. In a recent blog titled #EnoughAlready,[13] he also raises the concerning trend of social media platforms being used to vent and complain about other veterinary professionals. He directly addresses the need to state that the time has come to own our part in being responsible for our communications and words and provides guidelines to "to put an end to the cyberbullying and hate within our own profession".

In response to the increase in concerns directed at both individuals and practices, the AVMA created a "Cyberbullying and how to handle it" page and The Social DVM site also created a cyberbullying course. Additionally, 24/7 support for AVMA members through a crisis management hotline was developed as of December 2016. The Bernstein Crisis Management experts provide up to 30 minutes of free consultation initially followed by significantly discounted additional services if needed. (Bernstein Crisis Management @ 626-531-1140). Managing an online reputation can also be implemented (AVMA members again qualify for significant discounts) through Your Review Genius and DVM Reputation Guard. If your license is threatened, resources through AVMA's Professional Liability are available. Many countries now have resources for their veterinary community sharing "best practices" and support tools for social media use and navigating cyberbullying and harassment on their veterinary organization websites.

There is now quite a bit of advice available to reduce risks to individuals and businesses from social media. Dr Jason Coe, professor at the Dept. of Population Medicine at University of Guelph's Ontario Veterinary College and a leading expert in veterinary clinical communications, was honored in March 2021 by the AVMA with the Bustad Companion Animal Veterinarian of the Year Award. His research and teachings have made important and impactful contributions to the profession around many aspects of professional communication, including social media use and cyberbullying. Some of these can be seen in the AVMA's very useful cyberbullying information and resource hub at https://avma.org/resources-tools/practice-management/reputation/cyberbullying.

Telemedicine—Impacts on the Veterinary–Client–Patient Relationship and on Veterinary Practice

Our historic veterinary care model is evolving to one that needs to adapt to the current data- and technology-driven economy and consumer-driven expectations. Generational shifts in demographics as well as the adoption of telemedicine for human healthcare is pushing advances in what is now called "connected care" in veterinary medicine. Pet owners are seeking increased access to virtual patient care, ease of communication with medical staff, and convenience when it comes to booking appointments and requesting prescription refills. Connected care is, and will, impact all segments of our profession (companion animal, equine, food animal, zoo, etc.)

Virtual platforms are being created to provide private, secure access to patient medical records (for clients and veterinary care facilities). There are new/developing legal concerns regarding confidentiality of medical information and records being shared virtually. This is a growing discussion about the potential ethics regarding data usage, privacy, and record accessibility and sharing as the concept of HIPAA-level privacy is new for veterinary medicine. In an article by Dr Kerri Marshall, she evaluated challenges and ethical concerns surrounding veterinary connected care. She pointed to the current dynamic of veterinary professionals shifting from "hands-on-heroes" to "digital guides". She was homing in on the impact on the veterinary relationship with the client and the patient with the need to uphold our veterinary oath and remain trusted advisors on animal care. "The veterinary profession has dedicated itself to caring for animals and public health through education and training to be the key animal health advisors in the future".[14]

Some other issues that are at the center of the connected care ethics discussion are:

- Access to care/
- Veterinary practice efficiency and team's quality of life.
- Quality of care—standard of care.
- Medical record-sharing.
- Data privacy, security, accessibility to all care providers for the animal.
- Monetization of connected care.
- Prescribing and diagnostics.
- Crossing state or country boundaries.
- Access to pet medical records during the tele-veterinary consultation.

There is a moral imperative to move as a collaborative veterinary community toward designing digitally connected platforms as well as being involved in the creation of necessary laws and policies surrounding the use/access to technology. Currently, the AVMA, the Association for Veterinary Informatics, and the American Animal Hospital Association (AAHA) are collaborating to create policies and provide guidelines for practicing tele-veterinary medicine. No doubt this is just the beginning of such conversations, particularly after the 2020–2021 experiences in veterinary medicine where new protocols, technologies, and ways of working with our clients were created out of necessity to uphold COVID-19 safety measures.

The impacts and learnings from the COVID-19 pandemic of 2020–2021 on the veterinary profession, our larger communities, and ourselves are just starting to be understood. We use terms such as "curbside care" and PPE (personal protection equipment) now, as if these have always been a consideration. Despite vet workers in all segments adapting quickly out of necessity due to the shortage of people and the persistently high case numbers (estimated to be 40–60 percent higher than historical averages in some locations), the work over the last 16 months has exhausted teams around the world. A resounding theme heard in many discussions around the pandemic's impact on veterinary caregivers personally and professionally, is that it threw our preexisting challenges into stark relief which could no longer be ignored or untended. Our historical dysfunctions including the neglect of "people first", culturally respectful work environments, outdated technical adaptations, and stigma around mental health challenges are now in the forefront demanding change. The intensity, tenacity, and broad impacts of the type of burnout that developed this year as a result of the collective traumatic experience is resulting in a collective awakening and commitment to the needs of our veterinary community's wellbeing. Many articles, blogs, and research endeavors have been born from this time period that are exploring what is occurring for us as individuals and as teams and how we can innovate and evolve for the better of all and support the quality of care that we all wish to provide to animals. Out of this crisis, there will be post-traumatic growth and new creativity which may have much-needed positive impacts contributing to the improved holistic health of the veterinary industry.

The opportunity exists for there to be improved client experience if done in such a way that utilizes technology to increase convenience, communication, and efficiency, while upholding the client's perceived value of veterinary care provided. Equally, it will be vital to foster the mutual trust, respect, and civility between the veterinary staff and clients. This increased access to care, convenience, and ease of communication may support involvement of animal care providers throughout the entire life of an animal. Utilizing telecommunication platforms will also provide increased opportunities for education and for proactive care that may decrease illness, emergency hospital visits, and economic euthanasia, e.g., providing links to educational content curated by the veterinarians (and possibly even produced by the veterinary team at that practice!).

Altogether, if thoughtfully created and implemented, these technologies could improve the patient's quality of life, support the human–animal bond, maintain—and even fortify—the relationship between the veterinary care team and the client, and allow for veterinary practices to be economically successful. These tenets are in line with the current movement for "Fear Free/Low Stress" handling of veterinary patients as telemedicine consults may provide a much healthier approach (mentally and physically) for some animals. Importantly, all these factors may then contribute to decreased moral stressors, ethical conflicts, and instances of bullying between clients and veterinary caregivers whether in person or virtually in the form of cyberbullying.

FINANCIAL CONCERNS IMPACTING VETERINARY WELLBEING

In this section, I want to first address financial considerations for veterinarians (such as student debt and salary) and then I will discuss how client financial limitations can impact standard of care and the doctor–client–patient relationship.

Finances is one of the primary reasons for veterinarians to feel more stressed, to experience burnout, and to consider leaving the profession. Financial issues are a significant source of moral stressors, ethical conflict, and emotional distress. This is a multifaceted subject impacting veterinary medicine. My hope is that by evaluating the important ways finances impact our wellbeing and capacity to experience compassion satisfaction, we as a professional community will commit to having more intentional conversations and seek meaningful solutions. I particularly want to find the means to better prepare our young veterinarians to navigate these complex and emotional situations and to feel supported, knowing where to find the resources they need when they need them.

There are many important points to keep in mind when we consider the impact of finances on veterinary professionals and wellbeing. Depending upon where you live in the world, there are widely varying sociologic relationships to money and to what you earn in your profession. We are not going to delve into the socioeconomic inequities at this point, but they most certainly impact many individuals within the veterinary profession and our clients. The Access to Veterinary Care Coalition (AVCC) at the University of Tennessee was formed in 2016 to bring awareness to the lack of access to veterinary care in many communities, many times due to limited finances and the cost of veterinary care. In their 2018 report, AVCC reported that an estimated 29 million dogs and cats live in families participating in the Supplemental Nutrition Assistance Program (SNAP), and millions more are in financially struggling middle-class households.[15]

Broader societal views absolutely can contribute further to a feeling of either security or of scarcity. For example, the culture of the United States is one of high consumerism and capitalism. It is not unusual for people to spend what they make rather than finding ways to save or invest these monies. There are psychological impacts of working and living in a culture where "worth" and "happiness" are often defined by materialistic wealth. Who you are is defined by what you "do" and "what you earn" in many Western cultures versus what you contribute to society at large and how you improve the quality of life for all (humans, animals, and environment).

Income Potential vs. Debt

Regardless of where you practice in the world, generally many caregiving professional roles pay modestly. For most, the opportunity to move outside of a low-to-moderate earning range is rare. In the helping professions, the "pay" is more about satisfaction from the work itself and in the positive difference that is being made in patients' lives. Most veterinary associates and clinicians went into this profession because they wanted the opportunity to work with animals and they cared about animal welfare. We did not enter this profession solely for prestige or for high pay. Most are aware that there is a significant discrepancy between the earning potential for human caregivers and veterinary professionals, despite very similar large investments of time and money in our degrees. Those of us that chose this passion-driven professional path may or may not have had a clear understanding that most roles in the veterinary profession provide modest earnings. Veterinary caregivers are some of the most selfless, compassionate individuals. That does not, however, mean that we should not continue to seek a space where we can afford to live comfortably and to feel less distressed by debt and other financial pressures.

The financial concerns associated with the veterinary field are some of the biggest contributors to individuals feeling distressed, unfulfilled, compassion fatigue, and burnout. According to a 2019 article in The Washington Post that compassionately chronicled the story of a local veterinarian who contemplated suicide, finances played a significant role in her mental health challenges. The veterinarian shared that it was her feelings of overwhelm about her debt in relationship to her earning potential which played a significant role in why she had contemplated suicide. Fortunately, her close relationship and sense of obligation to her toy poodle helped her to change her mind and seek professional mental health support.[16]

Within this same article, important findings from the 2019 US Center for Disease Control (CDC) report on veterinary mortality rates in America were also touched upon. It was this report that first revealed the awful result that between 1979 and 2015, veterinarians died by suicide 2–3.5 times more frequently than the national average (females were more likely than males to contemplate suicide). The factors that contribute to suicide are complex and multifaceted. It is also very difficult to know what exactly leads a particular individual to die by suicide. However, for those that have contemplated suicide and have shared their stories, there is a trend that distress caused by finances is an important contributor. Veterinary tuition is expensive, more so if you choose (or must) pay out-of-state tuition. Per the AVMA, the average veterinary student now graduates with $143,000 or more in debt (overall cost of education for four years (tuition + cost of living)) with a national average of $147 K in post-graduation loans. Research shows that one-third of new graduating US DVMs have debt loads more than $200,000. There is a smaller percentage that have debt that exceeds $400,000! The negative impact of these debt burdens on mental health and on creating future financial security cannot be understated.

The average veterinary salary in the US continues to be low for a new graduate with an average around $67,000. According to Indeed and salary.com, the average veterinary technician salary in the US in 2021 is $36,018 (range $30,315–$42,798). There are variances in salary depending upon where in the US you are practicing, how many years have been spent in the profession, and the specific type of veterinary medicine that is being practiced. More recent reports of the average salary for a small animal practitioner are closer to $90,000 per year (as of 2020). If a veterinarian can move toward business ownership or pursue further training and certification in a specialty field, their earning potential may increase. Ownership opportunities are less and less common and creating the necessary capital to start one's own veterinary business is difficult if already saddled with significant debt. It would certainly take courage and the support of excellent business/financial advisors. Specialization is competitive and physically/emotionally/mentally grueling as a professional path. There is a commitment to three (or more) years of low pay, an intense schedule with little time off, and no guarantee of passing the necessary board examination, finding a great position, and having the desired increase in financial security and better schedule to support work–life harmony.

Most veterinarians in the US and in other countries, however, find that financial circumstances are a persistent source of concern and stress, more so if they have significant debt. Both veterinarians and veterinary technicians find themselves having to work additional hours, if not second jobs, to make ends meet or to save money to buy a home or fund any other investment. The debt/income disparity for veterinary technicians and clinicians with low incomes and high debt adversely impact wellbeing and professional fulfillment. These professionals find themselves earning lower-than-expected incomes for the investment of both time and money while earning their degrees to provide veterinary care. Additionally, historically there has been a low ceiling on earning capacity for many general practitioners, nursing staff, and practice managers.

Dr Dean Scott:

After veterinary school (in the 1990s), I had about $70 K in loans and the best job I could find offered $28.5 K, working about 50–60 hours/week, no benefits, and one week of vacation after the first year. Which is why I initially joined the Army Veterinary Corps!

The most recent Merck Animal Health Veterinarian Wellbeing Study demonstrated a trend between 2017 and 2019 that approximately 42 percent of respondents would not recommend a career in veterinary medicine. The high student debt, the relative low pay/salary in relation to debt, and the overall stress of working in veterinary medicine were the top reasons for not recommending veterinary medicine as a professional path.[17]

The concerns around debt because of rising tuition costs is a complex and difficult topic. In the last ten years, there has been a shift of financial burden of veterinary education from federal and state support to increased individual student tuition. There has concurrently been the addition of new private veterinary schools and increased class size. The tuition a student may pay is impacted by several factors:

1. Is the student paying in-state resident or non-resident tuition?
2. Where has the student been accepted and determined to pursue their veterinary education (US vs. Canada vs. island schools vs. international accredited institutions, e.g., UK, Ireland, Australia, New Zealand)?
3. Was financial aid an option for the student and to what degree? (e.g., option to accept commitment to military or to state residency/work after vet school graduation).

Interestingly, in speaking to veterinary colleagues and reviewing the data from the veterinary professional organizations in Australia, Canada, and the UK, the degree to which their veterinary students suffer the consequences of these tuition increases are relatively similar when compared to the US. In both the US and abroad, international students inevitably pay more and receive less financial assistance. The domestic, full fee-paying students abroad pay only slightly less than the international students, and this would be consistent with the majority of the US veterinary students. A small percentage of individuals abroad and in the US are supported more substantially by the government (e.g., commonwealth-supported students in Australia or USDA-sponsored Veterinary Medicine Loan Repayment Program). Also, the structure of their undergraduate/graduate training time is different than in the US, often with a 5–6-year total time investment. An article from the April 2020 Canadian Veterinary Journal examined the average school-related debt and income data for recent graduates. The respondents represented ten out of the 13 Canadian provinces and the majority were in small animal practice (86.5 percent) with a smaller percentage reporting from the equine, poultry, or production animal medicine segments (13.5 percent). The survey results indicated that the average annual base salary practicing in Canada is $73,241 (USD $57,502). The evaluation of school-related debt for the six Canadian veterinary schools 2017–2019 demonstrated that the average amount decreased for three of the veterinary colleges (FMV, UCVM, and WCVM) and increased for 2 (AVC, OVC). The range of average school-related debt in 2019 was $22,333–$143,353.[18]

Recently a British veterinary colleague shared that university fees are on average £9000 per year. Most students get a loan for this, but that does not include loans for living expenses ("maintenance loans"). She shared that her own maintenance loan in 2015 was about £3000/year. Taking those numbers into consideration, a five-year veterinary student would be looking at £60,000. Cambridge University is a six-year standard program (six-year BVetMed course with integrated BSc), and there are some students that do a "pre-vet year" before the five-year course at the Royal College, which means that their debt would be higher. In fact, a 2021 graduate from Bristol Veterinary College, Holly Stringfellow, shared the following with me:

> It's really hard to quantify exactly how much debt each person will have, but all will have at least £45,000 just from tuition, not including living costs, unless they are self-funded (https://www.gov.uk/student-finance/new-fulltime-students?step-by-step-nav =18045f76-ac04-41b7-b147-5687d8fbb64a). This may help in some regards—it all varies depending on household income as to how much student finance you're eligible for. And then of course there are the interest rates which will start to increase that burden as soon as we've graduated! If a student is receiving the maximum student finance loan—and not living in London—then they could be adding on an additional £46,000 (give or take as the number allowed changes each year). So already that's over £90,000 at the end of the five years.[19]

In 2015, the British Veterinary Association surveyed some of their younger members who had graduated from different UK universities. The survey sought to gauge the impact of student debt on career decisions and overall professional outlook.[20] Those surveyed were asked if they were aware of the amount of debt they would incur during their studies and whether having debt post-graduation caused them anxiety or affected their career choices. The case studies revealed that both the undergraduates and postgraduates felt similarly: that their debt did not cause them great worry and that they felt that their student debt was both manageable and justifiable. Importantly, the loans that many students receive are from the government, not private. The student loans are repaid directly from an individual's salary at a rate proportional to their annual income and accrue interest at a nominal rate. If private bank loans or credit cards were used to also fund the tuition fees, these require regular payments commencing immediately, or shortly after graduation. It will depend upon the individual's debt burden and the type of loans used as to whether they might feel pressure to get a job immediately after graduation or whether they may be able to afford pursuing additional training (internship +/- residency). There are an increasing number of four-year post-grad courses in the UK that may help those pursuing veterinary medicine to better manage the tuition debt burden.

In speaking to some of my Australian veterinary colleagues in GP, specialty, and in academia, the tuition fees, student debt, and government support around loans is very similar to that of both Canada and the UK. Average income levels were reported by them subjectively (their own as well as some of their professional colleagues in their community) and were found to be similar relative to the UK average income levels.[21]

At the organization/institution level, there are ongoing, robust conversations around managing class size, tuition, and providing more reasonable financial support for students (scholarships and veterinary corporation-supported loans). There remain, unfortunately, few debt-forgiveness and management options at present for veterinary professionals. For the profession to continue to draw young people of diverse backgrounds to consider veterinary medicine as a career path, tuition must be made more affordable and post-graduation financial stress diminished. These are absolutely opportunities for the veterinary community to collectively and creatively work toward viable solutions. I personally am choosing to be a realistic optimist about this subject and sincerely hope that global academic institutions share the concern about veterinary graduates experiencing less financial distress for the sake of the profession at large.

It will take time to find a way to support veterinary teaching colleges to function well and afford reasonable salaries for their staff while keeping tuition affordable. In the meantime, increasing knowledge around financial matters can significantly reduce the risk of these concerns contributing to serious psychological distress. Increasingly, Veterinary schools have added business and financial planning content into the curriculum over the years, which certainly is helpful. Creating financial understanding to allow for informed action toward future security is key. This allows for increased individual freedom to make life choices best for personal and professional reasons, not because of feeling "trapped" by one's financial situation. Both the AVMA and the VIN Foundation provide robust resources on calculating student debt: income, education on financial planning, and on determining best approaches to navigating student debt payment options post-graduation.

Student debt has also impacted individual decision-making about whether to pursue additional training following veterinary school (internship +/- residency path toward specialty medicine). In speaking to students, while I was in clinical practice and subsequently while on the Clinician Talent Acquisition team, I heard repeatedly that they could not afford the additional 1–4 years of training. In the last several years, there has been a distinct decrease in rotating internship applications through the Veterinary Internship and Residency Matching Program (VIRMP). As a result, there are also fewer applicants for special internships and residency programs. This is a concerning trend for our profession overall and for future of specialty veterinary care. I also, however, find it upsetting for a young, passionate veterinarian to feel that their professional aspirations must be limited due to financial insecurity. This anxiety around earning potential following graduation also impacts the decision of what type of veterinary medicine to practice and what job opportunities to entertain. Rather than pursuing the field of choice and choosing a position on best cultural and mentoring fit, many young applicants select their first job based on promised compensation structure (particularly the base + production models vs. salary). Finances are not separate from a practice or from an individual practitioner—reflective of behavior and values. Good medicine is good business. An individual practitioner's ability to make that a reality will be determined both by their first years of clinical practice and by the practice culture where they work, supporting the development of solid, positive clinical acumen. In the right environment, a veterinarian comes to appreciate the value of good medicine being provided and their own self-worth as a clinician providing this care.

Most veterinary professionals may not have considered engaging a financial planner if they are even in a position to afford one. Evidence demonstrates that having a "coach" to help you navigate your debt and your investments for increased financial security can contribute to the sense of "flourishing" overall rather than just "getting by". Perhaps the coach comes in the form of reading books, blogs, and articles that support a clearer understanding of budgeting, debt management, and investing. I personally recall a period where I was financially strained and recognized that part of the solution included me getting clear on what was in my control to do. I read, I listened, I reached out to trusted financially savvy individuals in my life to help me. It did not change the fact that there was no easy fix and that hard work and living frugally were needed, but I absolutely felt less anxious about my circumstances. I also came to understand that to be more financially successful, I would also need to address not knowing how to take a stand for myself financially. This is a widespread issue apparently where many veterinary professionals are financially unassertive when it comes to declaring one's worth. It is a worthwhile investment of your time (see what I did there?) to explore the resources that are available to veterinary professionals, particularly those that are familiar with navigating student debt concerns. I provided an initial list of resources in the appendix of this book to get you started.

Dealing with Clients and Money

This leads us into the next very important part of the discussion of stress to veterinary professionals around financial issues: working with clients and pet owners. There are so many aspects to identify and explore around this topic that we likely could dedicate an entire chapter to it. Importantly, like in other places in this book, the focus here is to take a high-level approach to allow for identification of concerns and a normalization of the distress that can occur when navigating these emotionally fraught conversations. Regardless of how long you have been practicing, where you work, or what field of veterinary medicine you are supporting, financial discussions with clients can be like walking through a minefield of potential moral stressors and ethical conflicts for us.

First and foremost, there are the unrealistic financial expectations from clients. Most are genuinely ignorant to the cost of medical care, as human insurance billing systems are highly variable with little transparency. As a result, value for amount of money requested/spent is a challenging topic. Clients also have no real concept of the cost of running a veterinary practice, of providing diagnostics, and of paying for qualified support staff and doctors to run the hospital. There are other factors to consider here as well that complicate these necessary conversations. In addition to the skewed perception of the "appropriate" costs and value for the expense, there is also a genuine gap of affordability of treatment with the provision of high-quality veterinary care.

Research efforts, as well as conversations with colleagues, demonstrate that there are millions of pets globally that do not receive even basic veterinary care due to cost, not simply ignorance of the clients. Access to veterinary care to support improved animal welfare and quality of life is a real issue, and there are only a few resources available in each country to support families in supporting improved animal husbandry (small and large animal alike). Shelter medicine, low-cost clinics, vaccination drives, volunteer efforts, organization/government initiatives absolutely matter and will continue to be important contributors to bettering animal care. Shelter medicine has now become its own specialty!

Financial constraints or a client's decision to not invest monies in recommended animal care for other reasons are huge sources of both ethical conflict and moral stress for veterinarians. There are often employer or management policies in place as well that may limit the flexibility of a veterinary care team in making treatment decisions. When you layer on the feelings of ethical and professional obligation that veterinary caregivers have, which may conflict with the available financial means, or the cultural conflict where clients see animals more as property than family members, you can see how distressing these situations can be for everyone involved. When a recently rescued puppy comes in for profuse vomiting and diarrhea resulting in profound dehydration, there is the clear need for some form of supportive medical care. In an ideal world, diagnostics would be done to narrow the differential diagnosis and to guide therapy. Regardless, the puppy would benefit from parenteral fluid therapy, antiemetics, analgesia, and possibly antibiotic therapy. I cannot tell you how many times I was faced with this sort of case only to have the family share one or all of the following:

1. They do not have the financial resources to do more than outpatient therapy (no diagnostics)—and limited treatments at that. They do not yet have an established relationship with a primary veterinarian so are not sure what they are going to do the next day for follow-up.
2. They just got the puppy and do not want to invest in an animal that "was probably already sick" or has something wrong with it that they do not want to be responsible for (they discuss wanting to return it for a "healthy" puppy).
3. It's Christmas and they just got the puppy for their kids, and they spent all their discretionary income on holiday presents resulting in their credit cards being maxed out. There are no additional financial resources to now pay for medical care for the puppy AND your practice does not allow for any free therapeutics to be provided to the puppy (e.g., subcutaneous fluid support and an injection of Cerenia).

I admit to feeling nauseated myself just writing out these scenarios and reliving some of the moral distress I felt when faced with these circumstances (many, many times). I am sure that you can relate.

Compassion fatigue?
No. I suffer from
idiot fatigue.

In emergent situations, the high emotional state of clients combined with the issues above can lead to a volatile mix. There is the need to quickly build trust and rapport with clients to educate them on the emergent concern for their animal and the potential financial commitment for appropriate and necessary treatment. The discussion around finances in an ER environment needs to also take into consideration that these were unforeseen expenses and often the care needed overall is more costly. To navigate these scenarios, patience, calm confidence, and compassionate, trauma-informed communication skills are all helpful. I see this as one of the areas for opportunity to develop both awareness and efficacy while veterinary professionals are still in training. Practice via role-playing these scenarios while in school could be hugely beneficial to everyone involved. When in the clinic environment post-graduation, continuing to have peer-mentorship to help younger veterinarians and technicians to be more proficient and successful is key—developing skill will take time and experience.

We are not done here. There are other potentially morally distressing scenarios that need to be considered. Not being able to provide care that was thought to be appropriate and necessary impacts the entire care team's emotional wellbeing and morale. Repeated instances may lead to a sense of burnout contributing to increased attrition when these caregivers cannot do what they were trained to do. On top of this, if the veterinarian and/or technicians are asked to do and/or provide treatment or procedures that were outside of the caregiver's skill set for financial reasons (or other reasons but were stated to be financial by the owner), this can cause severe distress to the caregiving team. "Economic euthanasia" lands in this area as well. Inappropriate requests for the procedure from clients or having to relieve suffering of an animal because truly the family cannot afford appropriate care/therapy, has the potential to cause deep and residual emotional and psychological impacts on the veterinary team. Economic euthanasia instances increase in times of crisis, whether that be the 2008 financial recession, the global COVID-19 pandemic in 2020–21, or any natural disaster. The energy and time invested in working through these difficult conversations and scenarios takes a toll on the veterinary caregiver's empathy level. You can see why financial issues are one of the most important contributors to empathic distress which can contribute to both compassion fatigue and burnout for veterinary professionals. Financial limitations of clients have consistently remained the most common ethical challenge in surveys by Batchelor and McKeegan (2012), Kipperman, et. al. (2018), and Quain (2021) and have been cited in many other studies as one of the most significant stressors overall for veterinary caregivers.[22,23]

Clients "pushing" boundaries and saying manipulatively hurtful things to veterinarians when it comes to money is simply awful. When you have been practicing long enough, you have heard clients state that "if you really cared about my pet's wellbeing, you would do this for less money (or for free!)". In some instances, the pet owner is desperate and so resorts to this emotional blackmail. More rarely, someone who believes that you can be manipulated to do more for less money if made to feel guilty will try this approach. You will hear that veterinarians and their practices are "in it for the money", not for the animal. The righteous posture of how they are the ones being manipulated cuts like a knife to our soft, compassionate hearts. And they know it. There are repercussions to "comping" services just to "be nice". Not only might we reinforce the unethical and manipulative behavior of the client, but the impact on the practice's business is important to consider. Without fairly charging for services rendered, there is less money available for staff hiring and raises as well as for upgraded equipment and therapeutics that will benefit other patients' care.

Having to justify and defend what we do and why we do it feels terrible. Owners arguing or "wheeling and dealing" when estimates are reviewed, is exhausting. Based on my discussions with colleagues and my own personal experience, it is hard not to feel resentful and to take it personally in many of these circumstances. Believe me, after 20 years and hundreds of these conversations with clients in the ER practice environment, I get it! We will not go into it here, but it is also demoralizing when professional colleagues tell clients and accuse others in their veterinary community of being dishonest for a profit or of price-gauging. Gross and yes, I have experienced this in my career and heard about it from too many of my professional peers over the decades. As a colleague pointed out, it is in our code of conduct not to disparage colleagues or to bring the profession into disrepute. Just saying.

EUTHANASIA AS A PART OF OUR VETERINARY OATH TO RELIEVE ANIMAL SUFFERING

Humane euthanasia is a responsibility and an opportunity to provide a compassionate, dignified end to an animal's suffering. As part of our veterinary oath and code of professional ethics in veterinary medicine, euthanasia is a valuable aspect of how we contribute to animal welfare at an individual patient level but also at a public health and community level. We know this at an intellectual level but that does not mean that performing euthanasia is not emotionally challenging. There are absolutely scenarios where humane euthanasia feels like the right solution for an animal and for the client/family and, although sad/disappointing/frustrating for a variety of reasons, when done well and kindly, the animal is respectfully released from their physical embodiment. Even when the above is true, having to euthanize patient after patient multiple times in a shift or in a short period of time is hard on the psyche.

In emergency medicine and in busy practice environments, when there is the need to quickly move from the emotional heaviness of a euthanasia or patient death for any reason to the next case with a refocused, open-hearted space, know that this is understandably challenging. The emotional burden is temporarily tucked away for later processing and integration to allow the caregivers to keep moving and caring for the other patients and families in need. I just want to take a moment to honor how courageous and selfless this is for our veterinary teams. It is also hard work and is considered some of the emotional labor that we do as part of our jobs—and it contributes to our feelings of fatigue. Healthy compartmentalization takes awareness, self-compassion, and practice but can be an essential element to sustaining mental health when coping with such potentially distressing emotions.

While considering potentially distressing scenarios around euthanasia, I want to discuss the unique and particularly difficult situation of "economic" euthanasia further. There are truly times where pet owners are not in a financial situation to provide the care needed to relieve an animal's suffering. Then there is the parallel scenario of "convenience" euthanasia. The pet owner requests that an animal be euthanized not because the animal is suffering but because they have become a "problem" or an "inconvenience". Rather than working through behavior issues or seeking to re-home an animal, these owners seek to euthanize the animal. Ethical conflict? You bet. Working through these conversations is tough and requires a heap of assertiveness, self-compassion, and clarity on one's values and ethical boundaries. This is also an opportunity to utilize narrative medicine and associated communication skills and listening. Take time to get the full story from the clients before being too quick to judge. There may be other factors in their lives or in the animal's history that you may also need to be clear on. Again, another great opportunity for veterinary caregivers and professionals in practice to talk about how to deal with this type of conversation before it actually happens and/or to debrief after they happen to allow for learning from one another.

The last scenario that I wanted to touch on with regard to euthanasia is when veterinarians must "cull" healthy animals to mitigate animal suffering or due to a public health concern. A 2011 article addressed an important new variant of traumatic stress that is defined as "perpetration-induced traumatic stress". In the veterinary profession, this mental stress and negative mental health outcome can be the result of both euthanizing healthy animals as well as having to participate in the euthanizing of large groups of animals to ease their suffering (e.g., hoarding situations or natural crisis), and animal management crisis, or as part of an epidemiological control of disease spread.[24]

It may be that a single animal is euthanized in an effort to control rabies in a particular community or several animals in a herd or flock if feared to be carrying a highly contagious disease that could impact other animals' or human health. These are tough enough, but I wonder how USDA veterinarians or international public-health veterinarians involved in zoonotic disease control cope when having to euthanize/cull/dispatch large numbers of animals at once to mitigate the spread of an infectious agent between animals and/or to humans. There have been instances in the recent past with millions of poultry in countries around the world having to be culled to control avian influenza. In 2020–2021, millions of minks on fur farms around the world have been "dispatched" to control the spread of SARS CoV-2 back and forth between human caregivers and the animals. During the pandemic, there were also thousands of pigs and other livestock that had to be euthanized because the COVID shutdown protocols led to disruption of the business chain—animals could not be transported due to closure of slaughterhouses. Other crises, such as the Australian bushfires of 2019–2020 and in the Western US in summer 2020, led to a significant increase in the number of animals (livestock and wildlife) that had to be euthanized.

I sincerely wonder (and hope) that there is support for the psychological health of veterinarians and technicians that must be involved in these circumstances. This may be another important arena of future opportunity for veterinary professionals to consider: how to protect public health while also respecting animals' quality of life and right to life? Public health and conversations around responsible farming overall are hugely important ethical conversations for veterinary professionals to be involved in as policy, protocols, and public awareness/education can all be positively impacted.

These challenging scenarios that we have just examined lead us to our next important discussion: the impact of secondary and vicarious trauma on empathic caregivers.

Trauma as Part of Our Caregiving Landscape

Caregivers navigate traumatizing situations and language in their work every day. When I bring up the concept of trauma being present in veterinary medicine, what comes to mind for you? I will bet the first thing would be the physical and possibly psychological trauma sustained by our veterinary patients. Although their trauma is indeed one of the factors that may result in the secondary trauma that we experience as veterinary practitioners, that is not the trauma I am referring to here. On rare occasions, when their fear and/or pain is overwhelming, we may indeed experience primary trauma when these patients physically cause harm (bite, kick, scratch, etc.).

What I am alluding to is the complex landscape of bringing historical trauma to our work lives and then, while in these roles, being subjected to secondary and vicarious trauma on a regular basis. It shows up in the form of clients coping with financial constraints and having to choose humane euthanasia over treatment for a pet. It is in the form of an animal vocalizing or demonstrating physical symptoms of distress. We see, feel, hear the pain of both the humans and the animals that are suffering, and we often tend to deeply empathize with them. In my own experience and in conversations with many veterinary colleagues, we know that we are particularly sensitive to the suffering of others, and it makes us deeply uncomfortable. Yet, we chose a profession that deals with that suffering up close and regularly. Recognizing that we intentionally chose this profession, and we choose the work that we each do every day to decrease suffering, even while we navigate the effects of empathic distress and secondary trauma on ourselves, is the epitome of generosity and compassion.

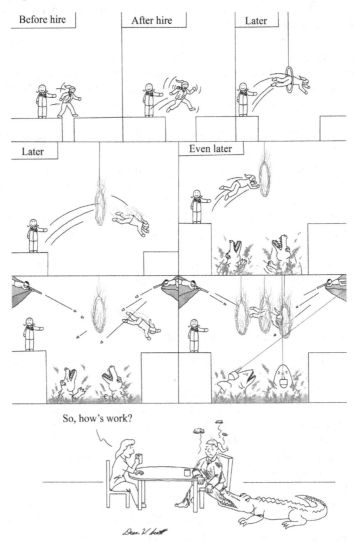

This concept of trauma-informed care is a new one to me and likely to many in our veterinary caregiving community. In preparation for both the book writing and my current role of clinician wellbeing trainer, I started with reading the works of the founding trauma-informed care researchers Drs Bessel Van Der Kolk, Peter Levine, Gabor Mate, and several others who have built upon their fundamental work. It is fascinating to contemplate the many ways that trauma is woven into the systems, culture, and individual workers that are in the helping professions. A full discussion will be beyond the scope of this book. However, the learnings are so valuable and invoke some important reflection that I strongly recommend that you consider taking time to engage in some of these learnings for yourself. I put a few choice texts in the recommended readings at the end of this book.

One of the books and authors that I found myself deeply resonating with is Dr Laura Van Dernoot Lipsky as I read her book, *Trauma Stewardship*.[25] I felt connected to her personal experience. Her approach to sharing her story, as well as those of other caregivers that she interviewed, was authentic and raw but expertly framed with clinical knowledge. I felt that she was sitting across a table and talking directly to and about me and my professional caregiving experience as an ER clinician. Dr Van Dernoot Lipsky writes with personal knowledge and understanding of her own experiences with primary and secondary trauma that led to compassion fatigue and eventually burnout. She takes a deep dive into what is known in western medicine around working with trauma and approaches how to compassionately navigate the emotional, physical, and mental minefield of this caring work.

In the reshaping of our professional culture toward one that supports "people first" language and cultural humility, this complex but vital construct needs to be incorporated. The wonderful advent of supporting "fear-free" caregiving environments for our veterinary patients is a good analogy. We are seeking to not worsen the experience for the patient and, if at all possible, to contribute to a more positive outcome for the veterinary patient, the client, and the caregivers. Trauma-informed care (TIC) and language also comes from having new knowledge, perspectives, language, and skills that can decrease unintended harm and promote more positive outcomes.

Trauma-Informed Care

So, what is "trauma-informed care"? The work of trauma-informed care began as a formalized movement in 2005, building momentum over the last 16 years. Trauma-informed care has been recognized as a vital part of any and all human care delivery systems, including healthcare settings, hospitals, community mental health centers, homeless shelters, school settings, foster care agencies, detention centers, and domestic violence services. As the movement grows, it is not uncommon to encounter such services as trauma-informed yoga and trauma-informed mindfulness.

There is a general need for increased awareness of how trauma manifests with our clients, our veterinary patients, and ourselves so we can be more thoughtful and compassionate to not make any given situation worse for anyone involved ("First, do no harm".). When I bring up the concept of trauma being present in veterinary medicine, what comes to mind for you? I will bet the first thing would be the physical and possibly psychological trauma sustained by our veterinary patients. Although their trauma is indeed one of the factors that may result in the secondary trauma that we experience as veterinary practitioners, that is not the trauma I am referring to here. So, what are some examples of how trauma-informed care and language manifest in our practice environments? Two excellent scenarios were provided to me by veterinary social worker Susan Swendsen Harris, MSSW, LCSW, VSW (personal communication, 2021):

1. There is the need for a trauma-informed approach in our care settings when examining our team dynamics. When trauma is experienced by a team, or team members have an extensive history of their own personal trauma, there can be a tendency to recreate these scenarios inside their teams. One particularly common scenario is that when someone starts to feel in a "victim" role, they inadvertently scan for who may be the "perpetrator", or "threat". When we recognize these unhealthy trauma responses occurring, a trauma-informed approach is to realize what is happening and respond in a trauma-informed manner.
2. In much of the veterinary community of the past year, there is a focus on the "angry client". Even in the wording "angry client", we are using a label that is disenfranchising to the perspective of the people that are possibly experiencing a difficult moment in the life of their loved one. In human healthcare, trauma-informed practice encourages us to adopt the language and mindset that says, "this person who is experiencing...", to recognize the emotion that is being observed is tied to this instance or event, and not a sum total judgment on that individual.

Importantly, taking this approach of slowing down and acknowledging as a compassionate human that this other human is suffering takes practice, patience, and self-awareness. It is equally important that self-care and safety are also kept in mind for all involved. Taking a trauma-informed approach as a caregiver does not mean that it gives another the right to be verbally, emotionally, or physically abusive to you. From personal experience, I know that finding that compassionate space for someone who is challenging in the moment and at the same time respecting yourself and allowing your own feelings can be difficult. Again, self-compassion practices are so valuable here. We have to remember to go easy on ourselves as we practice learning these communication and listening skills.

The most important aspects to know at this point is that taking a trauma-informed approach to care involves the entire organization and team being involved in the assessment, training, and ongoing support of the components to this framework. The paradigm shift from the "way that we have always done things" or making someone "wrong" for what they are experiencing, rather than examining what is "happening to them", are huge elements that can positively impact our veterinary culture evolution. The TIC approach also includes considering all levels of promoting and demonstrating safety (physical, emotional, and psychological) and a culture of wellness and self-care for all involved. There is an increased awareness of veterinary professionals who may have been the victim of trauma at one point in their career becoming the perpetrator, such as may happen when a more seasoned caregiver is training novice professionals. This can happen at any point in the training curriculum by individuals that may have the intention of "toughening up" the newer colleague or preparing them for "the reality of practice".

**SHARED BY A VETERINARY COLLEAGUE ABOUT
HIS VETERINARY SCHOOL EXPERIENCE:**

I have too many stories from vet school of outright bullying, dismissiveness, and condescension from professors, clinicians, and residents—those, in theory, who are supposed to be teaching us. I was a fourth-year student in clinics. I was mopping up urine from another student's patient so that they could get it where it needed to be. The emergency clinician was walking by and said, "It's good to work on your secondary skills, in case the primary ones don't work out". I mean, honestly, I'm surprised homicide isn't higher in our profession as well.

Another strategy common in systems with a strong trauma-informed implementation is to keep constant vigilance on the self-care and wellness practices of the caregivers. It is understood that empathic caregivers will experience trauma, either primary or secondary, and that wellness programs, such as mental health days, debriefings, and connection-building exercises (such as walking groups) have a huge value in building resilience. If I have piqued your curiosity, please explore the National Counsel for Behavioral Health website (www.thenationalcouncil.org) or the SAMHSA National Center for Trauma Informed Care website (www.samhsa.gov) for additional resources and information.

In the meantime, Susan shared that we may do well to consider the *"4 Rs"* of any work that is done where trauma exists, and these are:

1. *Realize that trauma exists, and how it can affect people, groups, and settings.*
2. *Recognize the signs of trauma.*
3. *Respond to the trauma with intention, purpose, and*
4. *Resist re-traumatizing whenever possible.*

VETERINARY PROFESSIONALS—COMMON PERSONALITY TRAITS THAT CAN GET IN OUR WAY OF FLOURISHING

Research has proven that there are some ubiquitous personality traits and behavioral tendencies that show up in both human and veterinary medicine caregiving professions. I am betting that you have read, heard, or discussed with colleagues how the following four concerns show up and can be annoyingly self-defeating once we get into clinical practice:

1. Perfectionism.
2. Imposter syndrome.
3. Over-identification with our veterinary roles—our self-worth is tied up in our work identity.
4. People-pleasing tendencies vs. assertiveness (important in our ability to declare our worth, needs, and define healthy boundaries).

These very tendencies and traits tended to be selected for when working toward veterinary training programs and then subsequently reinforced in the classroom/training environments. There is a positive feedback loop with these high-achiever traits. We learned to lean into and become adapted to the pressure that we put upon ourselves to be high achievers and excel in academics, to cram in a wide variety of extracurricular activities to increase our competitiveness before and after vet school, and to always say "yes" to potential professional-building opportunities As a care-giving community, we strive, we "muscle it out", and we stoically tolerate the lack of boundaries that would protect our mental and physical health. As vital as they may have been in helping us to "get into and stay in the club", they can be detrimental to our holistic wellbeing and flourishing in our lives if we rely on these tendencies alone. Let's take a closer look at both the "upsides" of these traits and the potential concerns when we don't keep them in balance.

PERFECTIONISM AND "HYPER-ACHEIVER" PERSONALITIES IN VETERINARY MEDICINE

It was stated by Dr. Ian Reckless in the 2013 Avoiding Errors in Adult Medicine (1st Ed.) that:

> It is human to err, but the response to that error is what is important. Doctors who make mistakes may become better at their jobs as a result. They can, and do, go on to have successful and productive careers. The key is to reflect on errors and pay heed to any lessons that can be learnt.[26]

I am a recovering perfectionist. I feel like many of us would benefit from being in a "Perfectionists Anonymous" 12-step program. It is very common for veterinary professionals to struggle with both perfectionism and imposter syndrome, as the two often go hand-in-hand. Our high-achieving tendencies were part of what fortified us to be competitive candidates to get into and survive the rigors of veterinary school. Our technicians also often have these same tendencies. There is a difference between setting high goals and then working hard to bring about the best results that you can vs. having unrealistic expectations of perfect outcomes.

There are many excellent articles, podcasts, and books about perfectionism in high-achieving professionals. Here are some of the most important and applicable take away points for us as veterinary professionals to consider:

- Perfectionism is the creativity killer and the fertile ground where procrasti-nation thrives. A personal example here: In the writing of this book over the last two years, I have been putting a lot of pressure on myself to have it be a certain way as I prepare myself to be vulnerable with my professional colleagues. I admit to getting stuck many times as a result of this perfectionist tendency showing up as a negative inner critic.
- When we are afraid to be seen as less than perfect, we end up saying "yes" too frequently (to our colleagues, our bosses, our clients) which can lead to working yourself to the bone and end up sabotaging your best efforts. In your efforts to show up, you scatter your energy and focus. You may miss

the opportunity to really do great work in one or two things, rather than so-so work on a multitude of things.

- Shame is likely to be triggered in perfectionists who experience failure or make a mistake (may contribute to "imposter syndrome").
- Increased likelihood of feelings of isolation, decreased help-seeking, and increase in anxiety, neuroticism, depression, and burnout. (Note: In research articles from New Zealand, US, and the UK, these factors have been found to be highest in female veterinarians, particularly recent graduates, and are significant contributors to poor mental health).
- Perfectionist traits when linked to imposter syndrome may result in us being much harsher critics of ourselves when we fail to meet our own unrealistically high standards of ourselves.

One of my favorite authors, Brené Brown, speaks and writes in such authentic and helpful ways about shame, vulnerability, trust, and learning to love our messy human selves and each other. In her book, *The Gifts of Imperfection*, there are far too many "yummy" quotes and points to list here about embracing and accepting that we are works in progress. One of the best to sum it up, however, is: "You're imperfect, and you're wired for struggle, but you are worthy of love and belonging".[27]

Yes, that means that we are *all* imperfect and also so capable of learning and growing! You must make best decisions with the information that you have currently in this moment. The action steps that you then take are more important than the paralysis resulting from the fear of failure resulting from needing your effort being viewed as perfect. Literally, take it one day and one small step at a time. This allows the path to develop in an informed way as you walk toward the larger end goals but at least you are walking! There are a multitude of incredibly insightful and helpful resources around how to navigate being your perfect self vs. being your most authentic and excellent self. I encourage you to get curious and find the approach that works best for you!

IMPOSTER SYNDROME

I am betting that most of us have had experiences of self-doubt at some point in our lives, particularly when learning something or doing something new or unfamiliar. Think about how you felt when you started a new job or when you finally landed in that much coveted seat in the veterinary classroom surrounded by fellow high achievers! That passing experience of self-doubt is perfectly normal and understandable. However, when you move into the overly self-critical realm of experiencing fear that you will be found to be incompetent or are convinced that you do not belong where you are, these are more likely to be attributed to imposter syndrome. The negative and self-sabotaging emotions and thoughts that arise when we feel like a fraud are experienced by many high-achieving individuals at some point in their lives. There are some of us, however, that are more likely to experience it on a regular basis and be hamstrung in our capacity to reach for new opportunities or lean into unfamiliar roles or relationships because of it. Yes, looking at you, fellow veterinary professionals.

Imposter syndrome is defined as a pervasive feeling of self-doubt, insecurity, or fraudulence despite often overwhelming evidence to the contrary. This is not unique to veterinary professionals as it shows up for smart, successful individuals in all professions, aspects of life, and cultures. In 1978, the term was first used in an article about high-achieving women.[28] The authors defined imposter phenomenon as an individual experience of self-perceived intellectual phoniness. Since then, many articles and research endeavors have explored this as an issue that negatively impacts creativity and self-efficacy and contributes to neuroticism. Competitive environments can lay the groundwork for perfectionist tendencies and that contributes to imposter syndrome. High achievers are more predisposed to suffer from these concerns, but research has demonstrated that around 70 percent of adults may experience "imposterism" at least once in their lifetime.[29] An international study done in 2020 specifically evaluating the incidence of imposter syndrome in veterinarians found a high prevalence of 68 percent in practicing clinicians. Being female or having been in practice for less than five years increased the odds of having a higher score on their survey.[30]

So, why does this matter? Imposterism can prevent you from pursuing new opportunities for growth in both work and in your personal life. You may buy into a false narrative and lower your self-esteem which impacts your performance in so many aspects of your life, including relationships. When combined with perfectionism, a cycle of self-doubt, fear, and self-criticism can become a toxic energy vortex.

"Superwomen" and "supermen" may push themselves to work harder than those around them to prove that they are not imposters.

Tackling this "inner critic" requires awareness, self-compassion, and a recognition of how growth requires making some mistakes initially as you develop new skills. This is an amazing opportunity to work with mindfulness, engage in peer–peer mentorship, and have honest dialogue with trusted family/friends, all of which can help each one of us to confront and find greater success in dealing with imposter syndrome. Recognize that when the "should" voice shows up, that is a fast-track to guilting and judging yourself and others. Just a sampling of the "Top 10 Playlist" that I have heard from myself and so many of my colleagues over the years: "I should be strong enough", "I should be smart enough", "I should put the animal patient's wellbeing above everything else". Sounds familiar, right?

DR KATIE FORD—VETERINARY SURGEON, SPEAKER, AND IMPOSTER BUSTER (WWW.KATIEFORDVET.COM OR @KATIEFORDVET). DR FORD'S TOP 10 POINTS TO MOVE THROUGH IMPOSTER SYNDROME:

Imposter syndrome is common, and it is not because you are a fraud. Remember, thoughts are not facts and we don't have to believe every narrative that we listen to. Often when imposter syndrome thoughts appear, we need to realign with our expectations, practice self-compassion, and understand more about why it comes up—often at times of growth or achievement. This ten-step exercise can help to do this:

1. Visit worse-case scenario briefly and have a plan. Decrease fear, increase control; demystify it! "Failure" = opportunity + information -> Growth!

2. List 3 reasons WHY you are choosing to do it. Your authentic "why"—you need to know and connect to your values. Don't be afraid to change your mind—it's your path!

3. List 5 reasons why you DO deserve it. What are the facts? Write down 3–5 daily wins—train your brain to see the good and see how you are succeeding.

4. List 3–5 strengths you'll need and where you have used them before (ask trusted friends and family—and believe them!).

5. List 3+ actions that you'll take to be kinder to yourself at this time (stockpile kind thoughts to draw upon when circumstances become challenging for you).

6. Ask: "Where did I do something that I didn't initially think I could do?" Focus on a positive narrative of yourself to balance out the negative stories in your head.

7. Ask: "WHO and HOW would I choose to show up?" You can choose your own, authentic identity rather than having it created for you by perceptions of what others think you should be. Attitude and actions are in our control.

8. Visit the BEST case scenarios (and use visualization!)—Why not get excited? Increase energy and positivity toward the outcome that you want to experience.

9. Who could I ask for help with this? Plan to action it. Remember to have your growth supported as you need it—teachers, coaches, therapists, family, friends. Success is not always solo! You do not have to know everything.

10. Write down: "I am not an imposter. I am valuable, skilled, and worthy regardless of the outcome". Include any other positive affirmations that speak to you as well, e.g., "you are valuable now!"

I really need to work on my Imposter Syndrome. I think I could do better.

Oh, no, your Imposter Syndrome is way better than mine.

You can also reframe your thoughts. In an interview withValerie Young, author of *The Secret Thoughts of Successful Women* (2011), she reminds people that the only difference between someone who experiences imposter syndrome and someone who does not is how they respond to challenges. "People who don't feel like impostors are no more intelligent or competent or capable than the rest of us", Young says. "It's very good news, because it means we just have to learn to think like non-impostors". Learning to value constructive criticism, understanding that you may actually be slowing your team down when you don't ask for help, or remembering that the more you practice a skill, the better that you will get at it can all be useful ways to shift your self-criticism toward an open-mindedness of learning.

This combination certainly feels familiar to me in the clinical environment with professional colleagues and clients as well as with leadership in work-related meetings. Over time as I grew in my experience level as a clinician, my confidence that my medical perspective held merit was bolstered. I learned to speak up and have healthy, collegial discussions which allowed for better patient care and for learning opportunities for all involved in the conversation. The same was certainly true with client communications.

As a novice practitioner, my imposter syndrome interfered with my capacity to authentically come from a space of knowledge and confidence. This is so incredibly natural and normal for a novice professional, but it felt terrible. It took time and trial-and-error to find the most effective words/phrases as well as the tailored approach to each client over years of practice. I have always maintained a "beginner's mind" about communication and human–human interactions. While in practice, I was keen to come from a place of calm, competent compassion and make sure I was genuinely listening to what clients, technicians, fellow clinicians, and leadership had to say. This, too, took time to develop and absolutely is an ongoing opportunity for growth in my many other work-related and personal roles. This is honestly a callout to all veterinary professionals to consider, as it is in our Veterinary Oath that we commit to lifelong learning.

In the writing of this book, as an example, I have called my imposter syndrome "my obnoxious inner roommate". I see and hear her sharing her unsolicited opinions about "who am I to write this book!" I take a deep breath, with compassion thank her for trying to protect me and keep me from getting hurt. I then ask her to step aside and please shut up now as I have work to do! Seriously, this is the conversation, and I just keep repeating it as often as is necessary to get this book written. I am not an expert, nor am I seeking to be seen as one, when it comes to veterinary wellbeing. I am, however, someone who cares very deeply for myself and my fellow veterinary colleagues. As amazing, smart, compassionate people, we deserve to be happy and this includes mitigating unnecessary suffering that results from isolation, stigma, and simply not knowing more. If collectively we aspire to give ourselves and each other the time and energy we deserve to be healthy in all regards, we can focus on the creation of work environments that support that lifestyle. With this, more of our internal resources can intentionally be allocated and channeled toward flourishing in all aspects of our personal and professional lives.

PEOPLE-PLEASING (AND OVER-IDENTIFICATION WITH OUR ROLES)

People-pleasing is another important behavioral trait that so many of us in caregiving professions embody. Research of both human and, more recently, veterinary caregivers indicate that this trait likely developed early in our lives (younger than 10 years of age). We may have evolved this strategy to provide ourselves a sense of being worthy of love and of care. This trait ties in closely with both perfectionism and with the tendency to be a hyper-achiever.

Faking it all day wears me out.

For those that have strong people-pleasing tendencies, there may be an unfounded fear of being labeled "aggressive", "emotional", or "arrogant" rather than self-confident when speaking up. We may find that we hold back on sharing our thoughts or valuable feedback in a conversation as a result. There may also be a tendency to prioritize others' needs above our own. People-pleasing is a huge contributor to our struggle to draw healthy boundaries that include the investment of time and energy into self-care and personal development. We may find ourselves saying "yes" to too many people and requests to our detriment. It is important to pause and consider what the repercussions are of saying "yes" and why you are saying it in this instance. Simply slowing yourself down enough to consider whether you are reacting, whether you are feeding a need to be needed, or whether the "yes" is the right response for you and for the occasion will take practice (believe me, I understand!). Consider whether "no", or "not right now", or "let me circle back to you" might be healthier choices for all involved. I love a callout by a friend recently as she deliberates how to respond to a request: choose the response that may make you feel guilty rather than resentful. The guilt comes up because we genuinely have this desire to be available all the time, to help, to make things better. Resentment means that a boundary has been crossed or you feel unappreciated or undervalued. Food for thought, hmm?

Another manifestation of our desire to be people-pleasers is to provide support and to make people feel better. I have recently been introduced to the concept of "premature reassurance".[31] In a nutshell, we might provide the reassurance to a client "not to worry" or "everything will be okay" before finishing a thorough workup on a patient. In examining why caregivers may choose to do this, we do very much want to ease the client's mind and decrease distressing emotions for them. However, it could also be that it is a result of our own fear about how a client might react, our own discomfort with being in the midst of distressing emotions (our own and others'!), the need to fix (believing that you really can fix whatever the medical concern is for the patient even before you have a diagnosis); or perhaps the philosophy of positive thinking will beget positive results. Even if intentions are good, premature reassurance can undermine the trust that the client may have of veterinary professionals.

We now have greater clarity around some important and common personality traits in veterinary caregivers. We touched on how these may impede our professional growth, influence our personal and professional relationships, and impact our mental health and wellbeing. Importantly, these same traits may also be contributors to the persistent stigma around mental health challenges and seeking professional help in the veterinary profession.

BARRIERS TO HELP-SEEKING

There is a myriad of considerations around why individuals may choose to, or not to, seek out support when they are in psychological distress. This is a complex topic that involves self-awareness, education to support self-efficacy, self-stigma, social stigma, as well as concern for detrimental personal and professional consequences. Perfectionism and imposter syndrome also impact a veterinary caregiver's awareness of, or denial of, there being a health or medical concern that needs to be addressed.

Part of it is that there is a lack of self-awareness around what is "going on" with them and that there is thus no clear value to seeking out mental health support and resources. Many veterinary associates are not well-informed about emotional and mental health concerns that may impact them while in training or in practice. As a result, an individual may not be able to clearly discern why they are experiencing the thoughts and emotions that are present for them and not realize that they are also very likely not alone in what is occurring within them.

I was discussing this phenomenon with a professional colleague who is a licensed veterinary social worker and worked with an academic veterinary community for many years. She asked if I understood the social science concept of "pluralistic ignorance". I had not heard the term, but once she explained it, I understood perfectly what was meant. She explained this psychosocial phenomenon to me and shared an excellent research paper that delved into help-seeking behavior in the veterinary community. In the article, pluralistic ignorance was considered a contributor to the low help-seeking behavior noted in the profession.[32] A great example is if someone in a classroom has a question but does not ask it, thinking that they are the only one in the room who did not understand or may have this question. That is a large assumption. The greater likelihood is that there are many others that do indeed have the exact same question! The fear of looking stupid or of not belonging in that group contributes to someone staying silent. Can you see how this may also apply to when an individual is struggling with confusion, anxiety, or depression in the professional environment as well?

It is also plausible that individuals do not recognize their feelings or behavioral, physical, and/or psychological symptoms as manifestations of varying types of work-related distress. This applies to also not being able to appreciate that their teammates and professional colleagues may also be experiencing similar concerns. (e.g., mentors supporting their house officer trainees). As a result, these individuals may not talk to anyone about their problems because they do not feel that they need to seek support or realize it may be valuable. As I am writing this book and our veterinary teams have sustained 14 months of unprecedented, challenging circumstances both in their personal and professional lives, there are many that are now manifesting behavior that we are calling "angry burnout". They are burned-out from both their home/personal lives as well as the exhausting work conditions. These individuals may be impatient, short-tempered, and standoffish. If they were asked if they need help or support, the response may be: "I am not the one with the problem! I have got this figured out; it is the rest of the incompetent folks around me that need to pull themselves together". Although this colleague may not recognize it, they absolutely need support and time for self-care to breathe, rest, and get perspective again.

There may also be a concern around lack of confidentiality. Many may not feel comfortable asking for support/help because they feel that they will be judged by their leadership or peers or seen as less capable of doing their jobs well. Sometimes this is legitimate; as practices are close-knit environments, gossiping does occur, and leadership may or may not be equipped to skillfully navigate these conversations in a helpful way. Increased awareness through education to heighten individual and team awareness of mental health concerns in the caregiving environment may allow for increased and earlier recognition of there being a concern and an increased feeling of psychological safety where these concerns can be discussed. These measures would allow someone who may find themselves struggling to then recognize and articulate their need for support and trust that they will receive respectful and compassionate responses from their teammates and leadership.

Veterinary professionals largely do *not* choose to share when they are in distress until they feel that they have run out of options for helping themselves. Stigma around demonstrating any inability to "handle things" as well as a bunch of other "shoulds" that we believe make up a consummate, successful veterinarian may interfere with seeking support, whether with family, colleagues, or mental health professionals. This is another manifestation of the perfectionist tendencies found in many veterinary professionals. Veterinary culture propagates this concept starting while in veterinary school, and it often persists in both subtle and more overt forms such as shaming, judging, and feeling guilty when individuals enter practice.

Studies in both human and in veterinary health professionals have demonstrated that those experiencing mental health challenges are more likely to speak to a peer, friend, or family member. In a recent study by Moses et. al. (2018), veterinarians reported that, in coping with a distressing situation, 72 percent would talk with a partner, a friend or colleague; only 12 percent stated that they would seek professional help; 17–22 percent reported "doing nothing" when asked about strategies that they had used to manage a distressing situation and how they had coped.[33] An important consideration here is that these distressing situations may have been a result of ethical conflict or moral stressors which are contextual and, in many instances, are a result of other elements at play within the practice/organization's larger systems outside of the individual associate's control. Professional help can help the individual to navigate the distressing emotions that may arise in these situations, but that does not address environmental or cultural issues which may have contributed.

There are some other very important reasons that an individual may not seek professional help when faced with mental health concerns:

- The cost of care (how much is covered by their health insurance if they have it?).
- Being provided time off with privacy and without negative repercussions on training time.
- Historical but real stigma of being seen as "incapable" or "faulty" by others.
- Perceived concerns about lack of confidentiality.
- Fears of documentation of their mental health status, or worries that disclosure will affect their academic prowess or their career.[34]
- The feeling that professional help will not help. There are (2) scenarios here. "Help rejecting" behavior is not uncommon in caregivers and can be associated with long-standing, "self-defeating" personality traits: "Nothing will help" or "I am managing on my own". However, an individual may be motivated to change behavior, such as to seek support, if demonstrable results are believed to be possible. There are likely many people that believe professional mental health support will not make a difference, e.g., "How can talking it out change my situation?"

KNOWLEDGE IS POWER: CLARIFYING MENTAL HEALTH CHALLENGES FOR VETERINARY CAREGIVERS

Caregiving as a profession can be deeply fulfilling. There are a multitude of meaningful contributions that are made to those who are the direct recipients of the care provided and to the larger community and environments that they live within. The "shadow side" of caregiving, however, includes conditions that challenge the mental health and overall wellbeing of the caregiving professional. We examined some of these concerns for caregivers overall in Chapter 2, and now we will look more closely at the potential impacts on veterinary professionals specifically.

Empathic Distress in Veterinary Caregivers Compared to Other Caregivers

As a veterinary community, we tend to be deeply empathetic individuals. To varying degrees, we chose veterinary medicine and caring for animal patients because of our innate ability to identify with and understand another's situation, feelings, and motivations when it comes to the relationship between animals and their caregivers.

We also have high levels of compassion, defined as a deep awareness of the suffering of another, combined with a wish to relieve that suffering. This compassion is extended most certainly to our veterinary patients and often also to the patient's owner/guardian. For example, when you see a cat struggling to breathe because of asthma, you are acutely aware of the patient's distress, and you move quickly to provide comfort and supportive care. Meanwhile, the cat's owners are in an exam room, crying and confused. After appropriate medical measures are taken to provide initial relief to the cat, you then enter the examination room prepared to provide support, education, and comfort to the emotional clients.

Humane euthanasia is a very good example of when we can absolutely experience empathy for a pet owner. In the process of euthanasia, we can relate to their sadness and grief as we have likely experienced similar feelings for our own pets when they passed. Our ability to understand at an emotional as well as a cognitive level the pain that they are experiencing is appropriate empathy. It is what allows us to provide a dignified death to veterinary patients and to support pet owners in an authentic way.

Historically, when we have referred to compassion fatigue in human and in veterinary medicine, we are describing a predictable and normal consequence of being an empathetic individual working in a helping/caring field. We can call this an "occupational hazard" for caregivers. As stated in a 2002 paper by Mitchener and Ogilvie, compassion fatigue was considered to be "the problem of strongly empathetic veterinarians who become internally depleted as a result of repeated exposure to emotionally challenging events on the background of minimal self-care".[35]

Compassion fatigue describes a set of symptoms, not a disease per se. These are symptoms that may wax and wane over time. Compassion fatigue symptoms often accrue over time if not recognized and addressed. The impact can be significant on your overall professional fulfillment as well as your personal mental and emotional health. In 2006, Figley and Roop surveyed over 200 practices in the US and over a third of veterinarians scored high or extremely high in the high-risk category for compassion fatigue.[36] It was Dr Figley who defined compassion fatigue as "the cost of caring" and he, along with other social behavior researchers, have stated that compassion fatigue is quite similar to secondary traumatic stress with elements of burnout.[37]

DEFINITIONS FROM THE COMPASSION FATIGUE AWARENESS PROJECT (WWW.COMPASSIONFATIGUE.ORG):

Compassion fatigue is the over identification with another individual's emotional pain. It starts with stress from the intense desire to meet the needs of others and results in the complete inability to relate to colleagues or clients.

Burnout is the state of physical and emotional exhaustion. It occurs gradually over time but is more likely to occur in long-term high stress situations, such as having physically and emotionally draining jobs.

(See the sections "Compassion Fatigue" & "Burnout" in Chapter 2 for more information)

The concept of compassion satisfaction as compared to compassion fatigue will be described in much more detail in Chapter 5. However, a brief description of compassion satisfaction to demonstrate the counterbalance to fatigue is to comprehend all of the ways that relieving suffering, that providing quality care to patients, that educating and supporting our colleagues and our clients result in feelings of purposefulness, self-worth, and contribution to the greater good in the world. In a more recent paper out of Australia, approximately 30 percent of veterinary students were at high risk of burnout, 24 percent at high risk of secondary traumatic stress, and 21 percent reported low compassion satisfaction.[38]

Therefore, having a clearer understanding of these concepts is imperative. You can develop then a higher degree of self-awareness and create your own unique "self-care" toolbox to mitigate the negative impacts of these symptoms. We will start to build that toolbox in the next chapter. Importantly, compassion does not fatigue or come close to running out for us as veterinary caregivers. The reality is that the physical and psychological symptoms of exhaustion can occur from caregiving without reasonable self-care practices and healthy boundaries to protect and replenish our energy stores. Empathic distress is more of a concern in that when a caregiver keeps taking on and feeling (literally and figuratively) the pain and suffering of animal patients and people, the burnout that occurs can lead to indifference and apathy. The remedy? *Compassion practices*! More to come on this but this is an important and more recent consideration for both human and veterinary caregivers.

STRESS/DISTRESS

In the last chapter, we defined and reviewed "stress", "eustress", and "distress" as mental and emotional states that directly impact human physiology and holistic well-being. Stressors of many types occur in our professional caregiving lives and environments. A quick review of these different types of stress:

Acute stress vs. chronic stress

Acute—the crisis happens; you adapt to it and/or find new coping mechanisms.
Chronic—a "simmer" or a slow "creep up" of insidious stress; it has more of a tendency to wear us out.

Eustress—Appropriate physiological reaction that occurs in life-preserving circumstances to increase our likelihood of survival. When stress is identified and managed appropriately, it can be motivating and useful in our lives.
Distress—Unpleasant, causes disruptive emotions and thoughts, can lead to mental and physical concerns, and is perceived as being outside of one's coping capacities.

There are physical stressors created by the work that is done by veterinary caregivers, such as long periods of time spent standing, manual labor with larger animals, caring for fractious (scared) patients, and sustained noxious auditory stimuli, to name just a handful. There are certainly many more, and these stressors contribute to the overall experience of distress associated with different veterinary roles.

Investigations of the myriad of stressors in veterinary medicine have been the focus of numerous research efforts and surveys over the last 15 years that have elucidated some of the top, persistent negative stressors in the veterinary profession. This is not an all-encompassing list but a nod to some of the most common cited concerns: schedule issues (on-call expectations, long shifts, shifts that do not support personal life balance); lack of adequate support staff; finances; client expectations; personal health; relationships (at work and the impact of work on personal relationships); poor nutrition; sleep deprivation or insomnia.[39] There is no doubt that all of these concerns (and others) may contribute to an individual's experience of compassion fatigue and burnout leading to mental and physical states of distress and/or exhaustion. What is less frequently pointed out and discussed is the contributing distress from moral stressors and ethical conflict.

MORAL STRESSORS AND ETHICAL CONFLICT IN VETERINARY MEDICINE

It would be helpful to frame these next several topics by first clarifying the concepts of moral stressors, moral distress, ethical conflict, and ethical exhaustion in veterinary practice. Dr Elizabeth Strand (founder of the veterinary social work program at University of Tennessee) shared a brief, but helpful summary of moral distress in the 2018 AAHA *Veterinary Practice Wellbeing Guide*:

> Moral distress: when external factors prevent team members from doing what they feel is "right".
>
> - Examples in practice: conflict with team members or with clients/pet-owners about end-of-life decisions, pet's quality of life, and standard of care (treatment approaches).
> - Negative emotions such as anger, sadness, fear, guilt may arise.
> - Dealing with these scenarios and negative emotions repeated over time results in emotional labor which can then resultingly contribute to compassion fatigue".[40]

Stated another way, "moral distress involves a threat to one's moral integrity. Moral integrity is the sense of wholeness and self-worth that comes from having clearly defined values that are congruent with one's actions and perceptions".[41]

These are topics that have been explored to a much larger extent in the human medical field. Data specifically pertaining to veterinarians remains limited but has been growing. What is known thus far for both human and veterinary caregivers is that moral conflict is an important form of occupational stress altogether. When a caregiver suffers from moral conflict repeatedly, moral distress may develop. When moral distress contributes to compassion fatigue and burnout, there then may be negative impacts on the individual, but there can also be increased conflict between teammates, medical errors leading to decreased patient safety, and an overall increase in staff turnover due to significant professional dissatisfaction.

The reality is that most individuals caring for animals, whether in a shelter, zoo, aquarium, farm, or veterinary practice, have probably experienced moral distress or ethical conflict, but perhaps did not know what to call the deeply impactful, negative feeling that "this is not right, and I am a part of it". Why does this matter other than experiencing feelings we historically have described as "sadness", "frustration", or "anger"? These scenarios can undermine our feelings of integrity and of authenticity. When we are repeatedly exposed to these situations, particularly if we feel helpless to make a different choice or feel forced into behaving in a way that is out of alignment with our values, moral dilemmas and ethical conflict are important contributors to caregiver's experience of burnout. Ethical exhaustion sets in when the individual gets to the point of feeling that they should stop trying or stop caring about the outcome. Empathy for patients and clients wanes over time. Some might get to the point where they feel that they are "just going through the motions" because they are disengaged and disassociating from their professional experience. They may believe that their actions will not make a difference.

In a 2019 paper that specifically focused on evaluation of moral stressors, moral distress, and ethical conflict in the veterinary population, the authors raised the additional question as to whether mental health concerns combined with moral distress may increase attrition of veterinarians.[42] In other research and professional forums in 2018–2019, several recurring topics and themes emerged. Some of the most common moral stressors and ethical conflicts were a result of:

- Financial constraints of clients resulting in a large amount of time and energy spent in discussions where there may be a disagreement about care for the pet between the owner and the veterinarian providing the care.
- Disagreement with management policies or feeling pressure from an employer.
- "Inappropriate" requests for euthanasia (convenience euthanasia) or contending with euthanasia as the only means of relieving an animal's suffering due to a client's financial constraints (economic euthanasia).
- Conflict around the feeling of having to provide care to a veterinary patient as an obligation because of an animal's legal status as property. In their 2018 article, Dr Lisa Moses and her co-authors shared: "...this feeling of obligation may come from within the veterinary culture and professional ethics itself, perhaps reflecting a cultural conflict between pets as family members and as property".[43]
- Not being able to provide care that they thought was appropriate.
- Being asked to do, and/or providing, treatment/procedures that were outside of the veterinarian's skill set for financial or other reasons (>50 percent reported being asked to do so).

> Dr Anne Quain, an experienced small animal GP colleague also pointed out in a personal communication that "our skill set (as GP practitioners) is shrinking because more of us practice defensive medicine. So as the standard of care increases, we need to refer more, so what is outside of our skill set is growing".

Biases and other influences (e.g., cultural, religious, "rules" associated with a particular organization/practice) also need to be taken into consideration. The concept of sentience and of the intrinsic value of animals is evolving, and there are active conversations in many circles about how this will impact veterinary professionals in training and in practice. Laws regarding animal and client rights are changing as a result in tandem. A commitment to staying engaged and educated on these topics, and on how standards of practice and animal law evolve as a result, is paramount.

In a 2018 interview with the aforementioned Dr Lisa Moses, who has herself been in veterinary medicine for more than 25 years, the conversation focused on exploring the "why" of increased suicide rates in veterinary medicine. She cited the results of the JAVMA article that compiled 30 years of data demonstrating that US vets are 2.1 and 3.5 times more likely to take their own lives than members of the general population.[44] It was in her 2018 JVIM study and article that she also poignantly had several practitioners share their personal experience of moral distress and ethical conflict:

- Howard Krum (retired aquatic animal medicine veterinarian):

We basically have a fatal flaw in our veterinary oath: we take an oath to serve animal welfare.... But the problem is animals are owned by people" from zoos, to companies, to individual pet owners. The problem of moral distress has been "looming and growing" in the industry for decades—an insidiously erosive force that seems very likely to be linked to the industry's high suicide rates.

- Etienne Cote (cardiologist at University of PEI): "To some people the pet that I see as my patient is, in fact, their child. It has that value. And to others, that pet is replaceable by another pet...(this) contributes very substantially to moral tension".
- Jennifer Michaels (neurologist/neurosurgeon in Boston): "... vets are often targeted for charging for their services and told "if they really cared they would work for free". The job can be thankless in many ways—clients calling at inappropriate hours, demanding prescriptions, or refills for patients that she has not seen or treated recently.

Now That We Know, How Do We Grow? Opportunities to Better Navigate Moral Stressors and Ethical Conflicts in Veterinary Medicine

How do we continue to learn and adapt our ways of practice, our training content, our intra-professional dialogue around such deeply complex and contextual topics as moral and ethical dilemmas in veterinary medicine? Here are some of the thoughts from subject matter experts on this complicated question (Moses, et.al., 2018; Quain, et.al., 2021):

For the Global Veterinary Community
- Training in navigating ethical dilemmas (practical guidance and training in recognizing, naming, and navigating ethical conflict) as part of the veterinary professional education.
- Ethical issues don't evaporate at the time of decision-making; an individual can still be left feeling moral distress. Raising awareness and creating "safe spaces" for debriefing with veterinary colleagues and leaders, allowing for emotional regulation and integration are vital.
- Veterinarians rely heavily on each other to navigate ethical dilemmas and counteract moral distress. Numerous studies point to the most common initial resource was discussion with colleagues.
- Normalizing the need for self-care, e.g., boundaries allowing for work–life integration. This may help to decrease the impact of ethical conflict, but the counterpoint is that as vet team members prioritize self-care, there may be an increase in ethical conflict awareness/feelings.
- Recognizing, acknowledging, and labeling conflict and distress as "ethical" in nature are important first steps in promoting ethical awareness.
- Utilize existing and create new practical solutions that do not place all of the burden on individuals to fix this problem themselves (e.g., development of policies and educational resources).

- Increase ethical literacy to facilitate understanding and acknowledgment of the frequency and role that ethical conflict plays in the mental and physical toll of practice.
- Accept ethical conflict as an inherent part of veterinary medicine to allow more common and distressing situations in practice to be seen through the lens of ethics.
- Formation of ethics committees, discussion and support groups, and ethics consultation services (exist in human hospitals and starting to be developed in large veterinary teaching hospitals).

For the individual veterinary professional

Here are some words of wisdom imparted by Dr Strand in the *"Moral DE-Stress"* article in the AAHA *Veterinary Practice Wellbeing Guide*, which supports mindful evaluation to facilitate learning and integration:

- What situations contributed to the negative emotions?
- In reflecting upon those conversations and situations, what can be identified that an individual did well?
- What was learned? How could that learning be applied in the future?
- Move toward positive emotions and mental state utilizing gratitude, humor, teamwork/collaboration, pride, and increased confidence.

In summary, the impact of moral stressors and ethical dilemmas is important to understand and to be able to navigate during clinical decision-making and practice. I asked Dr Quain to share some take away points based on her many years immersed in exploring these concerns for veterinary professionals:

Ethically challenging situations, or situations where it isn't easy to work out the best or "least worst" course of action, are common in veterinary settings. They are a known source of what is called moral stress and can lead to moral distress and moral injury. My research suggests that we need to:

1) Recognize that ethically challenging situations take time to deal with: we need time to identify stakeholders (anyone impacted by an ethical decision), gather information (which may include sourcing literature, policy or legislation, and/or talking to experts), and consider options.
2) Provide a psychologically safe space for ethical discussion or debriefing (like morbidity and mortality rounds but for ethical decisions).
3) Develop workplace policies that are helpful to the team members on the floor who have to implement them (and that require appropriate consultation and regular review).
4) Recognize that ethical challenges are a common, inevitable element of practice and proactively undertake continuing professional development in this area.

My research has shown that in the face of ethical challenges, veterinary team members seek advice, assistance, and validation from colleagues. Therefore, it is important that all team members develop skills in discussing ethical challenges—for example, in applying ethical decision-making frameworks". (Personal communication, March 2021)

SUBSTANCE USE DISORDER

Veterinary medicine is rewarding on so many levels, but as we have identified, there are also numerous potential reasons for veterinary caregivers to seek relief from some of the inherent challenges of our work. No one is above the impulse to put distance between themselves and traumatizing experiences. As humans, it is perfectly normal to seek comfort when we experience physical and psychological distress. However, the effort to numb or to disassociate from distressing thoughts and feelings can lead to overindulgence in food, sex, alcohol, and illicit drugs. Unhealthy coping mechanisms can become unhealthy habits that can harmfully impact all aspects of our health and lives. Whether unhealthy or healthy strategies are applied, it is the associated dopamine and serotonin release and the positive impact on mood that is being sought.

Substance use disorder (SUD) refers to when a person's use of alcohol or another substance (drug) leads to health issues or problems at work, school, or home. When discussing alcohol and drugs, addiction is a potential sequela of persistent and severe substance abuse disorder. People with SUD may have distorted thinking, behavior, and impaired judgment. Importantly, there can be changes in the brain's structure and function that perpetuate cravings and alter an individual's decision-making, learning, memory, and behavioral control. This combination of chronic changes to the brain creates a bypass of the prefrontal cortex (the rational thinking part of the brain). This is why the "Just Say No" campaign in the 1980s failed. This is a complex brain disease. Genetics, other mental health concerns (e.g., depression), chronic pain, emotional distress, and environmental stress can be potential factors that may lead to an individual being more predisposed to having this disorder. A stressful lifestyle and low self-esteem are other common contributors.

AS SHARED BY DR PHIL RICHMOND, VETERINARY COLLEAGUE AND CHAMPION OF INCREASING UNDERSTANDING AND DECREASING STIGMA AROUND USE DISORDERS IN THE VETERINARY PROFESSION (PERSONAL COMMUNICATION, MAY 2021):

Criteria for use disorders:

1. Substance is often taken in larger amounts and/or over a longer period than the patient intended.
2. Persistent attempts or one or more unsuccessful efforts made to cut down or control substance use.
3. A great deal of time is spent in activities necessary to obtain the substance, use the substance, or recover from effects.
4. Craving or strong desire or urge to use the substance.
5. Recurrent substance use resulting in a failure to fulfill major role obligations at work, school, or home.

6. Continued substance use despite having persistent or recurrent social or interpersonal problems caused or exacerbated by the effects of the substance.

7. Important social, occupational, or recreational activities given up or reduced because of substance use.

8. Recurrent substance use in situations in which it is physically hazardous.

9. Substance use is continued despite knowledge of having a persistent or recurrent physical or psychological problem that is likely to have been caused or exacerbated by the substance.

10. Tolerance, as defined by either of the following:
    a. Markedly increased amounts of the substance in order to achieve intoxication or desired effect.
    b. Markedly diminished effect with continued use of the same amount.

11. Withdrawal, as manifested by either of the following:
    a. The characteristic withdrawal syndrome for the substance.
    b. The same (or a closely related) substance is taken to relieve or avoid withdrawal symptoms.

Those with alcohol or substance use disorders should not feel that they are alone. As has already been identified in human medicine, the veterinary community needs to know that SUD and addiction are recoverable illnesses. Individuals that seek support should be lauded as courageous and given the support that they need to receive the therapy needed without fear of long-lasting personal and professional repercussions. When given the opportunity, an individual struggling with SUD or addiction can work with appropriate programs and peer support groups toward recovery and can return to practicing with reasonable skill and safety. There is ongoing work currently to move our veterinary boards and community toward compassion and creating support paths while decreasing stigma, but there remains quite a bit of work to do in this realm.

When examining substance use disorder as a concern for veterinary professionals, Dr Richmond shared some other valuable points for consideration. He noted that many states have a physician health program (PHP) that evaluates, recommends treatment, and monitors medical professionals being treated with a myriad of health concerns, including alcohol and substance use disorders. In most, but not all, of these states, veterinarians are included in their scope. Alcohol and substance use disorder is a mental illness and is treated as such. The data demonstrates that the rate of recovery in medical professionals who undergo treatment and monitoring is very high. There is a recovery rate of about 78 percent. In human medicine, about 10–15 percent of physicians will have an alcohol or substance use disorder at some point in their career.[45] As caregivers subjected to many of the same stressors as in human medicine, it makes sense that there would likely be a similar percentage of veterinary professionals that struggle with SUD.

On the *avma.org* website, there is robust information and resources available for those seeking guidance for themselves or a coworker. The "State Wellbeing Programs for Veterinary Professionals" content was prefaced with this paragraph:

> The AVMA Division of State Advocacy researched state laws and regulations that authorize or establish wellbeing programs for licensed veterinarians. Some programs cover other licensed veterinary professionals, such as veterinary technicians. These laws and regulations vary in scope and mission; the programs can include peer assistance, professional recovery, dependency, impairment, and diversion. The research below includes confidentiality provisions that we found that apply to participants in such programs.

This is another important area for us as a veterinary community to choose compassion over judgment. It literally could save a colleague's life. I encourage you to explore this topic and the growing resources further for yourself to increase your personal understanding. You can then shift your perception to one of being prepared and informed to hold space for someone that may need your support during critical moments.

SUICIDAL IDEATION AND SUICIDE IN THE VETERINARY PROFESSION

Why are veterinary professionals suffering and attempting/dying by suicide at rates approximated at 2.4 times that of the general population around the world? These and other related statistics are referred to in such a wide variety of articles, lectures, and discussions that I suspect that most veterinary readers here are familiar with them. There is, however, a vast difference between being familiar with and being complacent about such a disturbing, and persistent, trend in the veterinary profession.

How many others are thinking of it? How many have tried, but did not die? How many are numbing themselves with alcohol, drugs, and food? Far too many. I did not get quite to that point, but I personally suffered physically, emotionally, and mentally to the point where I had to stop practicing clinical medicine in 2015. I say with typical veterinary *gallows* humor that I was on a fast-track to a heart attack. But I actually mean it when I say it. Working overnight shifts of 12–16 hours for 20 years led to adrenal fatigue and cardiovascular arrhythmias. I now recognize that all of these physiological manifestations combined with chronic sleep deprivation and the psychological impacts of burnout contributed to other significant impacts on my life. I did not have the emotional regulation capacity that I knew I was capable of; I had decreased energy available for my family and friends; and I felt too exhausted, inside and out, to know how to find a different, healthier professional path for myself. It took me having increased cardiac arrhythmias for me to take all of these concerns more seriously and to know that I had to live my life differently. The result of nonaction might literally have led to my premature death.

What We Know about Suicide in the Veterinary Profession

In 2014, the CDC shared the results of their first mental health assessment of the veterinary profession. Of the 11,000 veterinarians surveyed, 1 in 6 reported considering suicide and 1 in 10 had suffered severe psychological distress with only 50 percent seeking professional help. At that time, a rate of 1 percent reported having attempted suicide. The subsequent Merck Animal Health Veterinary Wellbeing Studies completed in 2018 and 2019 provided research on veterinary professionals and rates of suicide that were much more detailed in their questions allowing more broadly reflective and granular results. We now have the data to verify what we all were observing in our respective work environments: veterinary technicians and doctors are at a significantly higher risk of dying from suicide than the general population. Here are the specific conclusions of the 2019 Merck study:[46]

- Female veterinarians are 3.5x more likely to die of suicide than the general population.
- Male veterinary technicians are 5x more likely to die of suicide.
- Male veterinarians are 2.1x more likely to die of suicide.
- Female veterinary technicians are 2.3x more likely to die of suicide.

These statistics are referred to in a wide variety of articles, lectures, and discussions. There is a difference between being familiar with and being complacent about such a disturbing, and persistent, trend in the veterinary profession. As I write this, I know of seven veterinary colleagues that died by suicide this week in March 2021. Seven this week.

I genuinely hope that we all remain heartsick when we consider the degree of suffering experienced by many of our professional colleagues that may result in individuals becoming so isolated and hopeless that they consider taking their lives to relieve the pain. Rather, I would hope that we use this data to reflect the starting point before we collectively determine to fight the stigma of mental health concerns that can lead to suicidal ideation and perhaps to suicide.

Research efforts and discussions from other countries demonstrate similarly concerning results. For example, a 2010 research effort by Platt et. al. demonstrated that veterinary surgeons in the UK are at least three times as likely to die from suicide as members of the general population, with an elevated risk for male vets. The authors stated that four main factors were identified as strongly correlated:[47]

- Personality characteristics of individuals entering the profession confer vulnerability.
- Psychosocial factors during undergraduate training and in the workplace, such as long hours, client expectations, emotional exhaustion, complaints, mistakes, as well as higher levels of anxiety and depressive symptoms.
- Access to and knowledge of lethal means.
- Familiarity with animal euthanasia leading to more permissive attitudes to suicide.

More recent research has resulted in the last point being contested. In a 2014 publication that explored the role of performing euthanasia on veterinarian depression and suicide, the authors found that euthanasia had a slightly protective effect regarding depressed mood and suicide risk. This is a complex topic and beyond the scope of this chapter and book. However, the specific point that was made on this issue in the article was the following: "Euthanasia administration and euthanasia-related interactions (e.g., the counsel of grieving clients) may allow veterinarians the opportunity to observe the impact that death has upon loved ones. Grief-stricken clients may serve as a reminder to veterinarians of the potential loss their loved ones would experience if they were to (die by) suicide. Second, for an individual suffering depression, the finality and irrevocable nature of suicide or death is sometimes overlooked but the administration of euthanasia may remind veterinarians of the finality of death".[48]

In 2012, Dr Brian McErlean, a veterinary surgeon with the Australian Veterinary Association (AVA) for over 30 years was tasked by the Western Australian (WA) government and the AVA with developing a suicide prevention program for veterinarians. The suicide rate in that state was found to be approximately four times that of the general population. In his professional opinion and research findings, Dr McErlean and others felt that the primary factors that contributed to the high suicide rate in WA were:

Isolation (in all forms) + depression + access to lethal means. The thinking at that time was that the combination of these factors with the relationship that veterinary professionals have with euthanasia as a humane ending of life when the individual is suffering contributed to the suicide rate being so high for this population.

In Australia and US, there are also the recent studies that compared the death records for veterinarians vs. those of the general population. In Australia, the standardized mortality ratio of veterinarians was 1.92 relative to the general population (Milner, et. al. 2015)[49] and American retrospective evaluation of death records of 11,620 veterinarians (1979–2015) concluded that male veterinarians were 2.1 and female veterinarians were 3.5 times as likely to die of suicide compared to the general population (Tomasi, et. al., 2019).[50]

From a May 2020 AVA article addressing the suicide rate of veterinary professionals in Australia, there were quite a few healthy callouts about the efforts being made to raise awareness overall and provide support:

We have a free 24-hour phone counseling service, run regular wellness events around the country and offer Mental Health First Aid training, with the goal of having a staff member qualified in Mental Health First Aid in every practice in Australia", said Dr Vale. The AVA's Graduate Mentoring Program pairs newly-graduating veterinarians with an experienced colleague in another practice to provide support, while a new AVA student group has been launched to help prepare upcoming vets for entering practice.[51]

Resources and Compassionate Help for Individuals in Crisis

Suicide Prevention Hotline: 1-800-273-TALK (8255)
Crisis Text Hotline: Text "HOME" to 741 741
www.suicidestop.com (International resources available here)
Note: Please also refer to Chapter 6 where I shared many more resources
to provide support for a wide variety of mental health concerns.

Not One More Vet (NOMV)

Founded in 2014, following the suicide of the world-renowned veterinarian and wonderful human, Dr Sophia Yin, NOMV is a peer-to-peer online veterinary support group and has grown to support a variety of resources dedicated to some of the toughest concerns we navigate in veterinary medicine, e.g., cyberbullying and harassment. The site has evolved over the years to include support for both clinicians and our wonderful support staff. It is intended to be a respectful, responsible, and ethical space to hear and support our colleagues who may be struggling. Importantly, NOMV has been vitally important and positively impactful in helping to normalize the conversation and decrease the stigma around mental health concerns and suicide in our profession.

Mental Health First Aid (MHFA) Program

The Mental Health First Aid Program was created in 2001 in Australia and came to the United States in 2008 via the National Council for Behavioral Health. The program is now being taught and used in many more countries around the world to increase awareness and decrease stigma around mental health disorders. Utilizing research-based approaches and people-first communication, the goal is to connect with an individual when early signs and symptoms are evident associated with varying mental health challenges. This early intervention can decrease the likelihood of a mental health crisis developing, support self-efficacy, decrease feelings of isolation for the individual that is struggling, and, in connecting these individuals to appropriate professional resources, may prevent self-harm or suicide.

Decreasing stigma around mental health concerns, substance abuse, and suicide and providing compassionate, respectful communication skills are the primary goals for the MFHA program. In their latest iteration, creating a self-care plan for those experiencing mental health challenges and for the mental health first aid caregivers is now also a part of the training. The MHFA programs are tailored to their audience of learners from teens, adults, first-responders, military, as well as others. Learning how to be more confident and skillful around supporting our fellow humans that may be struggling seems like an important consideration in today's world at large, not just in our veterinary profession.

VetLife Health Support

At the Medical Mind Matters Conference in 2015, Dr David Bartram 2015 shared the creation of the UK-based VetLife Health Support (VetLife). This independent charity provides free and confidential support to the veterinary community through a help line, health support program, and financial support. The help line (0303 040 2551) is maintained by volunteer listeners and, more recently, a confidential e-mail service has been added. From October 2014 to September 2015, it received 515 calls or e-mails.

VetLife provides collaborative care for mental health disorders and interfaces with the RCVS Health Protocol, which aims to protect animals and the interests of the public by helping veterinary surgeons and veterinary nurses whose fitness to practice may be impaired because of adverse health.

ASK (Assess–Support–Know) Suicide Prevention Training

- In 2020, Banfield Pet Hospital developed and launched this free, virtual training in conjunction with VetFolio.
- A 30-minute, interactive e-learning that is specifically designed for veterinary professionals.
- Helps to recognize when someone is in emotional distress or suicidal.
- How to connect them to professional support.

QPR (Question–Persuade–Refer) Suicide Prevention Training

- Training helps to prepare others to help others when in times of emotional crisis.
- Myths vs. facts, talking points, suggested approaches to uncomfortable conversations.
- Self-medicating and numbing.
- Feeling overwhelmed and not seeking help/support (huge impact of negative stigma around mental health concerns in general but particularly evident in the veterinary profession).

- This training seeks to raise awareness of the unfortunately high rates of suicide in veterinary medicine and to bring attention to the fact that suicide is a significant concern in society at large.

THE FAMILY DOG

—Author Unknown

If you can start the day without caffeine or pep pills,
If you can be cheerful, ignoring aches and pains,
If you can resist complaining and boring people with your troubles,
If you can eat the same food every day and be grateful for it,
If you can understand when loved ones are too busy to give you time,
If you can overlook when people take things out on you when,
through no fault of yours, something goes wrong,
If you can take criticism and blame without resentment,
If you can face the world without lies and deceit,
If you can conquer tension without medical help,
If you can relax without liquor,
If you can sleep without the aid of drugs,
If you can do all these things,
Then you are probably the family dog.

NOTES

1. Hofmeyer, A, et al. (2020). Contesting the Term "Compassion Fatigue": Integrating Findings from Social Neuroscience and Self-Care Research. *Collegian,* 27, 232–7.
2. Squire, L, et al. (2021). "We All Need a Variety of Tools for Our Mental Wellbeing". *Veterinary Record Careers.* https://www.vetrecordjobs.com/myvetfuture/article/-mental-wellbeing-toolbox-lucy-squire/
3. Larkin, M. (2020) "Veterinarians Talk Racial Discrimination". *Journal of the American Veterinary Medical Association,* 256 (12), 1293–316.
4. Armitage-Chan, E, and May, SA. (2018). Identity, Environment, and Mental Wellbeing in the Veterinary Profession. *Veterinary Record*, 183 (2), 68.
5. Gardner, DH, and Hini, D. (2006). Work-Related Stress in the Veterinary Profession in New Zealand. *New Zealand Veterinary Journal,* 54, 119–24.
6. Gardner, DH, and Rasmussen W. (2018). Workplace Bullying and Relationships with Health and Performances among a Sample of New Zealand Veterinarians. *New Zealand Veterinary Journal*, 66, 57–63.
7. Church, SL. (2018). The Horse: Your Guide to Equine Health Care. *Intensive Care for Equine Veterinarians.* TheHorse.com/160433.
8. Volk, JO, et al. (2020). Executive Summary of the Merck Animal Health Veterinary Wellbeing Study II. *Journal of American Veterinary Medical Association,* 256 (11): 1237–44.

9. Mullan, S, and Fawcett (Quain), A. (2017). *Veterinary Ethics: Navigating Tough Cases.* Sheffield: 5M Publishing Ltd., 368–71.

10. American Veterinary Medical Association (AVMA) Resources & Tools. 2021. *Guidelines on Social Media Community.* https://www.avma.org/resources-tools

11. Mandlik, R, and Larkin, M. (2014). Fighting the Cyberbully: How Harassment Can Affect Your Practice. *Journal of American Veterinary Medical Association News.* https://www.avma.org/javma-news/2014-11-14/fighting-cyberbully

12. Winter, M. (2020). Stop Workplace Cyberbullying. *Today's Veterinary Nurse,* 66–9.

13. Garcia, ED. (2020) *#EnoughAlready.* Online publication associated with his professional website: https://ericgarciafl.com/enoughalready/

14. Marshall, K. (2019). "How the Digital Revolution Raises Ethical and Legal Questions in Veterinary Medicine". *Veterinary Practice News.* https://veterinarypracticenews.com/ethical-issues-for-todays-veterinarian-in-the-digital-age/3/

15. AVCC Report. (2018). Access to Veterinary Care (AAVC): Barriers, Current Practices, and Public Policy. University of Tennessee College of Social Work, 5. https://pphe.utk.edu/wp-content/uploads/2020/09/avcc-report.pdf

16. Leffler, D. (2019) Suicides among Veterinarians Become a Growing Problem. *The Washington Post.* (Health & Sciences Section, January 23, 2019).

17. Volk, JO, et al. Executive Summary of the Merck Animal Health Veterinary Wellbeing Study II. *Journal of American Veterinary Medical Association,* 256 (11): 1237–44.

18. Canadian Veterinary Medical Association (2020). Class of 2019 New Graduate Survey results. *Canadian Veterinary Journal,* 61, 362–3.

19. Personal communication with Dr Holly Stringfellow, June 2021.

20. Marr, B. (2015). Becoming a vet…Is It Worth the Student Debt? *Veterinary Record Careers,* 22 October. https://www.vetrecordjobs.com/myvetfuture/

21. Quain, A, and Hazel, S, in Separate Conversations January–February 2021.

22. Batchelor, CEM, and McKeegan, DEF. (2012) Survey of the Frequency and Perceived Stressfulness of Ethical Dilemmas Encountered in UK Veterinary Practice. *Vet Record,* 170, 19.

23. Kipperman, B., et al. (2018.) Ethical Dilemmas Encountered by Small Animal Veterinarians: Characterisation, Responses, Consequences, and Beliefs Regarding Euthanasia. *Vet Record,* 182, 548.

24. Whiting, TL, and Marion, CL. (2011). Perpetration-Induced Traumatic Stress—A Risk for Veterinarians Involved in the Destruction of Healthy Animals. *Canadian Veterinary Journal,* 52 (7), 794–6.

25. Van Dernoot Lipsky, L. (2009). *Trauma Stewardship: An Everyday Guide to Caring for Self While Caring for Others.* San Francisco, CA: Berrett-Koehler.

26. Reckless, I, et al. (2013). *Avoiding Errors in Adult Medicine.* Oxford: Wiley-Blackwell.

27. Brown, B. (2010) *The Gifts of Imperfection: Let Go of Who You Think You're Supposed to Be and Embrace Who You Are.* Minnesota: Hazelden Publishing.

28. Clance, PR, and Imes, SA. (1978). The Imposter Phenomenon in High-Achieving Women: Dynamics and Therapeutic Intervention. *Psychotherapy Theory, Research and Practice,* 15 (3), 1–8.

29. Birchall, E, and Cronkwright, S. (2021). Overcoming Imposter Syndrome: How to Stop Feeling Like a Fraud. *Psychology Today* (published online: https://www.psychologyto-day.com/us/blog/conquer-the-clutter/202105/overcoming-the-imposter).

30. Kogan, LR., et al. (2020). Veterinarians and Imposter Syndrome: An Exploratory Study. *Vet Record.* https://bvajournals.onlinelibrary.wiley.com/doi/10.1136/vr.105914

31. Adams, CL, and Kurtz, S. (2017). *Skills for Communicating in Veterinary Medicine,* 2nd ed. Oxford: Otmoor Publishing.

32. Karaffa, KM, and Hancock, T. (2019). Mental Health Stigma and Veterinary Medical Students' Attitudes Toward Seeking Professional Psychological Help. *Journal of Veterinary Medical Education.* 46 (4): 459–9.

33. Moses, L, et al. (2018). Ethical Conflict and Moral Distress in Veterinary Medicine: A Survey of North American Veterinarians. *Journal of Veterinary Internal Medicine*, 32 (6), 2115–22.

34. Knipe, D, et al. (2018). Mental Health in Medical, Dentistry, and Veterinary Students: Cross-Sectional Online Survey. *BJPsych Open*, 4, 441–6.

35. Mitchner, KL, and Ogilvie, GK. (2002). Understanding Compassion Fatigue: Keys for the Caring Veterinary Healthcare Team. *Journal of the American Veterinary Medical Association,* 28, 307–10.

36. Figley, CR, and Roop, RG. (2006). *Compassion Fatigue in the Animal-Care Community.* Washington, DC: Humane Society Press.

37. Figley, CR. (2013). *Compassion Fatigue: Coping with Secondary Traumatic Stress Disorder in those Who Treat the Traumatized.* CR Figley (ed). New York: Routledge.

38. McArthur, M, et al. (2017). Resilience in Veterinary Students and the Predictive Role of Mindfulness and Self-Compassion. *Journal of Veterinary Medicine Education*, 44 (1), 106–15.

39. Vande Griek, OH, et al. (2018). Development of a Taxonomy of Practice-Related Stressors Experienced by Veterinarians in the United States. *Journal of Veterinary Medical Association,* 252 (2), 227–33.

40. American Animal Health Association. (2018). AAHA's Guide to Veterinary Practice Wellbeing. https://ams.aaha.org

41. Hardingham, LB. (2004). Integrity and Moral Residue: Nurses as Participants in a Moral Community. *Nursing Philosophy*, 5 (3), 127–34.

42. Arbe Montoya, A, et al. (2019). Moral Distress in Veterinarians. *Vet Record,* 185 (20), 631.

43. Moses, L, et al. (2018). Ethical Conflict and Moral Distress in Veterinary Medicine: A Survey of North American Veterinarians. *Journal of Veterinary Internal Medicine*, 32 (6), 2115–22.

44. Nett, RJ, et al. (2015). Risk Factors for Suicide, Attitudes Toward Mental Illness, and Practice-Related Stressors among US Veterinarians. *Journal of American Veterinary Medical Association,* 247 (8), 945–55.

45. DuPont, RL, McLellan, AT, White, WL, Merlo, LJ, and Gold, MS. (2020). Setting the Standard for Recovery: Physicians' Health Programs. *Journal of Substance Abuse Treatment*, 36 (2), 159–71.

46. Volk, JO, et al. (2020). Executive Summary of the Merck Animal Health Veterinary Wellbeing Study II. *Journal of American Veterinary Medical Association,* 256 (11), 1237–44.

47. Platt, B, et al. (2012). Suicidal Behavior and Psychosocial Problems in Veterinary Surgeons: A Systemic Review. *Social Psychiatry and Psychiatric Epidemiology,* 47 (2), 223–40.

48. Tran, L, Crane, M, and Phillips, J. (2014). The Distinct Role of Performing Euthanasia on Depression and Suicide in Veterinarians. *Journal of Occupational Health Psychology,* 19 (2), 123–32.

49. Milner, AJ, et al. (2015). Suicide in Veterinarians and Veterinary Nurses in Australia 2001–2012. *Australian Veterinary Journal,* 93 (9), 308–10.

50. Tomasi, SE, et al. (2019). Suicide among Veterinarians in the United States from 1979 through 2015. *Journal of the American Veterinary Medical Association,* 254 (1), 104–12.

51. Australian Veterinary Association. (May 7, 2020). Understanding the Factors behind Vet Suicide. *VetVoice.* https://www.vetvoice.com/au

4 Tailoring Your Individual Toolbox for Self-Care and Resilience Development

Take the first step in faith. You don't have to see the whole staircase, just take the first step.

—Martin Luther King, Jr.

You can have a fulfilling career as a veterinary professional *AND* have a life that you love!

Does your inner critic shout that the above statement is "impossible" or perhaps "naïve?" Does your fear of the unknown and of change hold you back? I get it. Repeat after me: I am worth it *and* more than capable!

When you come out of veterinary school, and you are thrust both into the excitement but also into all the stressors of clinical practice, you may experience a sense of "what have I done?" You might be thinking, "I have spent all of this time, and all of this money, and I am not sure that I belong here!" or "What do I do if this is not the right career path for me?" That feels awful and, for so many, there may be a distinct feeling of being trapped by your life choices and the time/financial investments that led to this point.

The good news? You are intelligent and you have higher-than-normal critical thinking capabilities around problem-solving. When faced with a challenge, you have been trained your entire life to persevere and to strategize. Now is the time for you to apply these skills toward supporting your wellbeing, as you practice in this amazing profession where you will positively impact so many lives. You have much more control and self-efficacy than you may believe. However, it takes a shift in mindset, it means declaring yourself innately worthy, and it also means that you need to unapologetically create the construct of your life that allows for both compassionate care of yourself and for the things that matter most to you in your life.

As medical professionals, preventative medicine is a familiar construct as a frame of mind when considering care and health. It may be useful to apply it here to the conversation on our self-care and professional community wellbeing. We are trained in and understand the value of taking informed steps to prevent, or to mitigate, symptoms of an illness. In many regards, this approach along with considering the entirety of the organism, rather than focusing on one body system, can lead to increased holistic health and longevity. To be forewarned is to be forearmed. Preparation for known occupational hazards and for the unforeseen circumstances

DOI: 10.1201/9780367816766-4

of your personal and professional life puts control back in your hands. Being more present and intentional in the moment, while also developing skills to fortify your future self to handle future problems with less personal suffering, is the intention of this chapter.

Why is this important? There is a lot of unnecessary suffering in this world that comes from ignorance and from lack of education and communication around a wide variety of human-related issues. The opportunity to raise awareness of the more common concerns for veterinary professionals, that impact our wellbeing and the way that we choose to live our lives, puts each of us back in the driver's seat of our lives. Together, I believe that we can diminish so much of the human suffering we personally experience and that we see all around us in our veterinary workplaces.

In this chapter, we will review the principles and practices that can support sustainable wellbeing. These principles shift your mindset from sacrifice to the innate value you hold that deserves care and attention. The practices give you daily tools to take personal control and accountability for your holistic health. The suggested practices, ranging from sleep habits to mindful breathing and intentional establishment of healthy boundaries are intended to pique your curiosity as you discover what supports you best through practice.

When you harness these practices and tools in your professional and personal life, you will embody valuable truths: you are not alone, you are not deficient, you are resilient. By more skillfully navigating self-limiting and defeating thoughts and behaviors that are common among veterinary professionals, you will build self-awareness and increase your self-confidence. Importantly, you will have a greater sense of control over your personal and professional destiny.

CONSEQUENCES OF POOR HEALTH: "WHY DOES IT MATTER?"

The World Health Organization defines "mental health" as a "state of wellbeing in which every individual realizes his or her own potential, can cope with the normal stresses of life, can work productively and fruitfully, and is able to contribute to her or his community" (2013). This definition emphasizes the flourishing aspects of wellness, rather than viewing health merely as an absence of illness.

In prior chapters, we have referenced Dr Charles Figley, the well-known trauma specialist who is also considered the "founding father of compassion fatigue research". He, along with several other professional colleagues, created the Green Cross Academy of Traumatology in the mid-1990s. The Green Cross was formed initially in response to the trauma training and disaster relief needs associated with the Oklahoma City bombing of the Federal Building in 1995. The organization has continued and evolved since that crisis event. It is a valuable non-profit organization providing humanitarian aid as well as trauma and compassion fatigue education to first-responders. There are a number of excellent certification courses that the Green Cross provides.

An important piece of the Green Cross Compassion Fatigue (CF) education for caregivers is founded in the "ethical duty of the caregiver to commit to self-care".

This evidence-based statement points to historical research which demonstrates that insufficient self-care results in poor quality care for patients and clients. Specifically, the CF educator and CF therapist course materials state:

> There is a moral obligation of caregiving professionals to help themselves—for their sake as well as for their patient's sake. The oft used but apt concept of putting on your own oxygen mask before you can help others with their oxygen mask? Yes, that is exactly the point—you cannot do your best as a care provider if you are not genuinely caring for your own mental and physical health. Those that are certified through The Green Cross Academy of Traumatology to support other humans in crisis (including the other caregivers) has created a code of ethics regarding self-care:

- Unethical not to attend to self-care.
- Sufficient self-care prevents harming those we serve.
- Person's own responsibility to care for him- or herself.
- No situation or person can justify neglecting self-care.[1]

Note: The "Standards of Care Guidelines" are worth reading through as you assess and create YOUR unique self-care toolbox: https://greencross.org).

What are the negative impacts of *not* taking adequate time for self-care? Here are some examples from the literature:

Psychological Impacts

- "… stressed and anxious doctors do not provide quality care and are costly to organizations";[2] "(doctors reporting depressive symptoms have been found) to make more (self-reported) medication errors".[3]
- Veterinarians—self-reported practitioner stress, illness, and fatigue have been shown to be causes of medical error in veterinary practice.[4]
- Impact on an individual and on their family (taking work and associated stress home).

Physical Consequences

Based on a large prospective literature search on burnout studies performed on a wide variety of occupations including medical professionals (2000–2012),[5] physical consequences include:

- Burnout has been found to be a significant predictor of coronary heart disease and a contributor to risk factors for diseases, such as type 2 diabetes, musculoskeletal pain development, and respiratory issues.
- Burnout can increase the likelihood of psychological concerns (e.g., insomnia, anxiety, depression).
- Increased gastrointestinal issues, headaches, and fatigue were all significantly associated with burnout.

The costs have been well-documented, but there are solutions available. If we can change our mindset, we can focus on new principles that guide our wellbeing. We will now examine some of the strategies that may better support you than the traditional way many of us have been coping with the high costs of our professional lives. First, we need to embrace and learn to practice the core principles of self-compassion and self-care.

Self-Compassion—It Starts with You

> I analyze and doubt my past experiences and my future deliberations (and miss the present). This constant self-monitoring is wrapped in the belief that I'm not enough nor being enough (in reality, I'm often doing too much and I am always enough). The self-evaluation is relentless—there's barely enough room for self-love.
>
> *—Sebene Sebassie (You Belong, 2020)*

The foundation of mental health, of ethical caregiving, and of flourishing in our lives lies with self-compassion. Giving yourself permission to redirect that incredibly generous, giving, compassionate heart that you have toward yourself. This is not indulgent. This is necessary. Pause for a moment and recognize how incredibly hard you are on yourself regularly with the inner judgment and criticism. It is much more helpful to choose a growth mindset with both optimism and self-compassion

When it comes to the topic of self-compassion, psychologists Dr Kristen Neff and Dr Chris Germer have been focused on this throughout their careers. There are bountiful books, TedTalks, and articles that cite their insightful and impactful research efforts. One of the important simplifying constructs that I love from Dr

Neff is that in defining self-compassion, there are (3) important fundamental components to consider:

1. **Self-kindness**: caring for yourself as you recognize your innate worth and value.
2. **Common humanity**: connecting with the support and external resources of knowledge, skills, and perspective from your communities. Decrease feelings of isolation and increase feelings of purpose and resilience.
3. **Mindfulness**: as defined by one of the founders of secular mindfulness in the United States, Jon Kabat-Zin: mindfulness is the nonjudgmental, objective evaluation of what is occurring in the present moment.

With these three concepts in mind, we will spend the remainder of this chapter supporting you in fortifying your holistic wellbeing through compassionate care of yourself.

Self-Care is the Core of the Self-Kindness Realm

Self-care is not about eating salads, doing yoga, and sitting in bubble baths. Although, truth be told, those all sound lovely! Self-care is about considering the many elements that comprise holistic health in a way that makes sense for you. Only you know what you need, what works and does not work, and what is possible in your life at this time. With that in mind, consider that in many fields of research, the determination has been made to consider eight (8) realms of wellbeing in the "wellbeing wheel": physical, intellectual, spiritual, emotional, social, financial, occupational, and environmental. These are interdependent dimensions that are present for each of us whether we are focusing on them at that time or not. The opportunity is to reflect on these realms for yourself and determine where to place your energy at this time that will positively impact you today. The bonus is that any positive action intentionally taken in one domain, positively impacts the other domains. Here are some brief clarifications as provided from SAMSHA's focus on Dr Nikita Gupta 2019 UCLA GRIT Coaching Program (Source: https://www.samhsa.gov):

Social Wellness

- Pursuing satisfying relationships with others.
- Respecting differences of other groups and individuals.
- Engaging in effective ways of resolving conflicts.
- Contributing to the common welfare of the community.
- Recognizing that oneself and society are interdependent.
- Understanding of personal and social identity within the larger community.

Physical Wellness

- Pursuing healthy, safe practices in areas of exercise, sleep, nutrition, and sexuality.

- Engaging in self-care behavior which promotes thriving and prevents illness/disease.

Emotional Wellness

- Being aware of and accepting one's own feelings and the feelings of others.
- Experiencing self-esteem and appreciating one's life.
- Paying attention to, expressing, and managing one's emotions appropriately.
- Managing stress and dealing with difficult decisions effectively.

Career Wellness

- Preparing for and entering work that is consistent with one's personal interests and values.
- Gaining satisfaction from work that is personally enriching and rewarding.
- Expanding and evolving one's skills and interests throughout life.

Intellectual/Creative Wellness

- Engaging one's mind in creative and stimulating activities.
- Using resources to expand knowledge and improve skills.
- Adapting to changes, new information, differing perceptions and approaches.

Financial Wellness

- Obtaining, managing, and maintaining finances.
- Paying bills, allocating finances appropriately.
- Healthy relationship with money.

Environmental Wellness

- Acting with recognition of the interdependence of self, society, and the natural environment.
- Accepting personal and social responsibility for promoting ecological wellbeing.
- Making environmentally sound choices concerning the workplace, home, and neighborhood.
- Maintaining home/work/study space in a way that supports success and thriving.

Existential (Spiritual) Wellness

- Values toward meaning and purpose in life—outside of or within the context of religious tradition.

- Respecting life's progression and significance.
- Developing trust, integrity, accountability, and an ethical approach to life.
- Understanding personal and social identity (including ethnicity, sexual orientation, gender, etc.).

Here is a quick way to determine your current level of self-care:

Self-Care Assessment

Take note of the internal dialogue that may arise as you consider investing time and energy into yourself. Do you have any resistance to making yourself and your well-being a priority?

a. What are some areas of self-care that you invest more time/energy in and are more comfortable doing?
b. Are there areas of self-care that are being neglected?
c. What would you like to include more of in your life?
d. What do you need now? What time, resources, and space would support that form of self-care?

You can also use the wellness wheel to go through each domain of your life and determine what is energizing you and what is draining you. Solutions and insights may arise as you take the time to evaluate intentionally and compassionately what works for you, what does not, and what else you may need.

WELLBEING TOOLBOX BUILDING TIME!

Once the principles are in place, we want to have daily, effective practices to support our individual mental and physical health. The practices include sleep, developing healthy coping strategies, cognitive reframing, and building resilience. You are also going to want to understand how to establish and maintain boundaries to support work–life integration and appreciate how mindfulness practices can complement all of these other practices. The foundation to all of this starts with healing and restoration through quality sleep.

Sleep Hygiene

> ... wintering is at once a season of the natural world, a respite our bodies require, and a state of mind.

> —*Katherine May, (Wintering: The Power of Rest and Retreat in Difficult Times, 2020)*

There is growing evidence around "sleep science", which reveals that sleep is foundational to our physical health, our mental health, and our cognitive capabilities. Sleep deprivation has a multitude of concerns: it impairs attention and long-term memory, reduces empathy and emotional regulation capabilities, and exacerbates physical concerns associated with chronic stress, such as decreased immunity and increased endocrine dysfunction. The concept of *"sleep hygiene"* does not refer to the brushing of teeth or making sure your sheets are clean, rather it refers to the developing of healthy habits that promote (and protect) good sleep.

What is "good" sleep? Although there are individual variances, particularly in different stages of our lives, research has proven that we ideally need seven to nine hours of sleep in a 24-hour period. This allows for a full four to six full sleep cycles to be completed, which provides the maximum benefits of sleep to the brain and the body. Why is good quality sleep important? There are many health benefits, but the three that stand out as key are:

1. Refreshing mind and body.
2. Processing of knowledge and regulation of emotions.
3. Removing waste products from our brain ("cerebral flushing") and from other tissues in the body.

Disruptions to our circadian rhythm that can contribute to insomnia versus having a healthy, uninterrupted period of sleep can come in many forms. Here are some of the common ones:

1. Changes in light cycles.
2. Melatonin and dopamine levels vs. cortisol levels.
3. Dietary factors, e.g., magnesium deficiency, ingestion of alcohol or caffeine close to bedtime, eating within 2 hours of trying to sleep.
4. Emotional or mental stress.

It would be valuable to assess whether your current habits are supporting your need for good sleep or are more likely interfering with the quality and duration of sleep you specifically need. What would be some signs of "poor" sleep hygiene? Having a hard time falling asleep, experiencing frequent sleep disturbances, and suffering daytime sleepiness may be attributed to poor sleep hygiene. An overall lack of consistency in sleep quantity or quality can also be a symptom of poor sleep hygiene. Importantly, there are also underlying medical conditions that could be contributing to poor sleep overall. It would be valuable to have a conversation with your primary doctor if you find that adjusting your sleep hygiene is not relieving any of the above signs as other steps may need to be taken to support your fundamental physical and mental health.

For each of us, setting the stage for sleep will look a bit different. It might require a commitment to putting away work earlier in the evening (for those that sleep at night!), to take some space for yourselves, or to address your eating habits before you lay down to rest. Since the root causes of sleep issues are unique for each of us, so too will be our remedies. There are some fantastic apps that help support your creating healthy sleep habits as well as helping you to fall asleep. Whether it is music, nature sounds, guided meditations to ease you to sleep, a story or podcast intended to calm your mind, only you know in the moment what will work best for you. Like anything else, you will not know what works unless you try. Some of the apps that are available also provide the opportunity to track your sleep (they do this by monitoring your biorhythms, or you can journal directly into the app). This is helpful if you are determining more objectively what is impacting your particular sleep patterns. Please see the appendix for some recommendations on apps that are currently available to support "good" sleep.

In the meantime, here are some suggestions on specific techniques that may be of help:

- Writing out your "to do" list for the next day before bedtime (get it out of your head!).
- Spending one-hour before bedtime device-free (decreases blue light which lowers melatonin levels—your "sleep hormone").
- Reading or engaging in another quiet, self-soothing activity such as listening to music, knitting, meditation, snuggling with your pet(s), reading a story to your children.
- Softening through the senses. You might explore the use of essential oils, taking a warm bath (lavender bath bomb!), listening to tranquil music or nature sounds, drinking a warm cup of herbal tea (Note: passionflower herb tea can help to reduce anxiety before bedtime).
- Journaling to express and release any recurring thoughts or feelings. Use this as your "worry time" to help clear and quiet your mind (Consider timing it for 10–15 minutes maximum, so that you are intentional in your writing and not slipping into rumination).
- Take time for gratitude reflection: name three things that you are grateful for in that moment.

- Stretching or practicing yoga to release physical tension in your muscles, soothing sore joints, and quieting your brain waves. (e.g., yoga nidra is specifically designed to be calming to your entire body).

As someone that worked the overnight shift for over 20 years, I understand the unique challenges that shift work creates around healthful sleep routines. Simply bringing your awareness to what you do that is helpful and what is not, can help you to make individual choices that you can then adapt to your own unique routines. For my fellow shift workers, here are some things that I found over the years to be important considerations when you need to find rest during daylight hours:

- Should you invest in blackout curtains and darker/warmer colors in your room?
- Is the temperature of your room comfortable?
- Can you use an alarm that does not startle you awake?
- What are you ingesting (literally and psychologically) and when before you lay down to rest?
- Do you need to practice using an eye mask and/or ear plugs?
- Do you need to allow yourself a nap? "Waking rest" is another way to provide a boost of energy, focus, and cognitive processing. The Japanese nap break known as "inemuri" is literally translated as "sleeping while present". (Note: "Napping: Do's and Don'ts for Healthy Adults" https://www.mayo-clinic.org/healthy-lifestyle/adult-health/in-depth/napping/art-20048319)

COPING MECHANISMS AND STRATEGIES

Developing coping skills and strategies that uniquely suit you will take reflection, practice, and accountability. Only you know what you need and what specific approaches will provide you the replenishment and perspective that works in your body, brain, and world. As veterinary caregivers, there may be perceptions that deter us from more closely examining our coping mechanisms as healthy vs. unhealthy:

- Perception: "I 'should' have all the answers to problem-solve and 'fix' this ..."
- Perception: "I 'should', 'must', or 'have to' sort out my issues on my own or risk being judged (or shamed)".
- Perception: "I feel like seeking help is a sign of weakness. I will be stigmatized by my professional community if I share that I need mental health support".

Opportunity: Choose courage and vulnerability not only for your own sake but for those around you who will see your example.

Assessment of Your Current Coping Strategies

When assessing your current state of being, remember that whatever you are feeling is allowed. You have a right to feel what you may feel, but it is helpful to remember that *you* are not defined by these emotions. These are simply transient states of being. You may be *feeling* overwhelmed, but your entire self is not defined as *being* overwhelmed. I love the analogy of emotions, thoughts, and feelings being like weather systems moving through us and our self—our identity is the "eye of the storm".

With practice, you can bring greater attention and skill to celebrating when there are positive emotions present and you can better navigate through challenging and uncomfortable emotions.[6] Our natural human tendency, also known as the negativity bias, evolutionarily was present to keep us safe. In the modern world, these adaptive ways of thinking and of acting can be self-defeating. We can become trapped in the vortex of negative thinking. I have heard it said in many places and by many voices that our human brain is like Velcro to negative experiences and is like Teflon to positive experiences. Consider for yourself for a moment when someone gives you a compliment, do you pause, let the positive in, and say "thank you"? Or perhaps your inclination is to move quickly on and dismiss or "hand back" the compliment by saying something self-demeaning? An example would be when someone tells you that you handled a situation well at work and instead of simply taking a moment to acknowledge that you did do a great job of diffusing that client's anger and find a positive resolution to their concern, you perhaps are quick to judge the client and/or yourself in saying that it "could have gone better".

Note the self-judgment that arises when you reflect on your emotional state or on your feelings. Do you say to yourself that you "should" be feeling this or that and that these feelings are "good" and those are "bad?" Again, just an important callout, acknowledging our natural tendency to do this. It is our human, analytical brain making sense of our world and how we are experiencing it moment to moment.

> Opportunity: Demonstrate a moment of self-compassion. Breathe three deep breaths to center yourself and slow your thinking. You will find that you are not only calmer and more focused but are also better able to act and speak in a more intentional manner.

EVALUATING AND UPDATING YOUR "COPING TOOLBOX"

In evaluating your current coping techniques, consider whether they serve you today, are they supportive, or are they counterproductive or draining to the "you of today". Some coping strategies are obviously healthier than others but only you can determine what works for you and what you need. With that in mind, however, perhaps there is value to considering concepts for healthier strategies overall:

- **Self-Soothing**—utilizing your five senses to comfort and center yourself as well as intentional, slow breathing. An example is the S.T.O.P. practice for a centering, restorative moment:

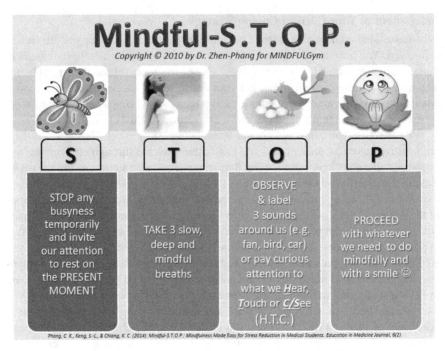

(Image courtesy of MINDFULGym (mindfulgymmalaysia.com) by Dr Phang Cheng Kar)

- **Distraction**—redirecting your attention onto another task and into other parts of your brain (e.g., puzzles, books, artwork, knitting, crossword puzzles, podcasts, TED Talks, music, movies, writing, other).
- **Opposite Action**—doing something the opposite of your impulse, which is consistent with a more positive emotion. Affirmations and inspirations (e.g., looking at or drawing motivational statements or images). Watch or read something funny or cheering (e.g., cat videos, funny movies, comics, TikTok silliness, or books).
- **Emotional Awareness**—developing tools for identifying and expressing your emotions (check out Plutchik's Wheel of Emotions and this article The Emotion Wheel: What It Is and How to Use It [+PDF] (positivepsychology.com)). Practice journaling or utilizing art to allow for interpretation of what you are emotionally experiencing and what may be triggering those emotions.
- **Mindfulness**—centering and grounding yourself in the present moment while letting go of any inner criticism and storytelling that may arise.
  Examples: Guided meditation (there are many wonderful apps such as Headspace, Calm, 10% Happier), soothing or uplifting music, grounding yourself with your senses (pet an animal, eat a piece of yummy chocolate, step outside and take a deep breath as you feel the temperature of the air, hold a meaningful object in your hand like a "worry stone"), practicing

mindful movement such as Qigong or yoga, mindful breathing exercises (e.g., "box breathing").

- **Crisis Plan**—Have contact information of your support network and of mental health support resources in your mobile phone. Make it easy to find and use these resources if ever needed when your own coping skills are not enough to bring you back to a place of balance and healthy action. You could also adopt Brené Brown's 1-inch square piece of paper in your wallet idea where you list all the names of your immediate, trustworthy support network that you could contact at any time if needed. What are the other resources that you have available to you through work (e.g., EAP or social workers)? What hotlines are you aware of, and do you have the contact information handy if ever needed by yourself or for a friend in distress? (Consider programming them into your phone!)

PRACTICE OF BOX BREATHING

1. Intentionally pause and either slow down if you are moving, sit quietly, or stretch out on the floor in a comfortable posture.
2. Allow yourself to settle and perhaps lower your gaze or, if it is alright and safe for you, close your eyes altogether.
3. Breathe in through your nose slowly for a count of four. Feel the air in your nostrils, throat, and entering your lungs. Note the temperature of the air.
4. Hold your breath for four seconds. Try to avoid exhaling and just rest in the momentary space that exists at the top of the inhalation.
5. Slowly exhale through your mouth for four seconds. Bring awareness to the feeling of the breath gently exiting your body and perhaps the sensation of your abdominal muscles and diaphragm assisting the exhalation.
6. Pause for four seconds at the base of the exhalation. Acknowledge the release from the exhalation and the anticipation of the next deep inhalation that exists in this moment between breaths.

Now that we have started to have a bit more awareness around where we are currently with our self-care and coping strategies, it is helpful to learn how to tap into the power of your mind. Specifically, developing skills that support a more positive mindset.

COGNITIVE REFRAMING

Thoughts are real but not true.[7]

—Tara Brach, **clinical psychologist and world-renowned mindful meditation author and teacher**

Reframing is an incredibly valuable practice of shifting mindset toward optimism and growth opportunity in whatever may arise in life. Change your attitude with gratitude. Simply taking the time to pause, come into what you are feeling and thinking, and consider what is the opportunity in this present moment for you. If you choose, you can shift from life circumstances happening *to* you to the understanding that life is *simply happening*. You are only a victim if you allow yourself to be in that mindset. The practice of taking a mental "step back" from a situation and choosing to be curiously objective can be incredibly powerful and empowering.

Pause for a moment and recognize how incredibly hard you are on yourself regularly with the inner judgment and criticism. It is much more helpful to choose a growth mindset with both optimism and self-compassion. With that brain space ready to receive information, I will share some of the strategies that better support wellbeing than the traditional way many of us have been coping with the high costs of our professional lives.

Utilizing Reframing to Transform Conflict

Conflict is a natural and integral part of our personal and professional lives. If we allow it to be, it can also be a formidable teacher. Conflict is an essential part of growth. The idea of conflict equating to some sort of personal failing is unfair to ourselves and also very likely not true. However, with that being said, we can learn to be more mindful in the ways that we speak and act so that we are contributing less to conflict occurring and propagating.

The ability to transform conflict, to sit with the uncomfortable feelings and emotionally regulate, and to set healthy boundaries are crucial components of wellbeing. Bring to mind and be true to your purpose, values, likes, and dislikes. Of equal importance is to identify the negative self-talk and judgment that may arise in the face of conflict. Slowing down and investigating with curiosity the body sensations and the automatic thoughts that arise for you in the midst of conflict will provide the platform to move through the "distress" and shift from the reactionary amygdala and limbic systems into the "upstairs brain", the neocortex.

Practices:

- *If you are feeling pessimistic*: try preparing mentally each morning and bolster your self-esteem.
- *If you are feeling alone*: identify your support network and help them to help you.
- *If you are feeling overwhelmed*: try letting go and working smarter, not harder.
- *If you are feeling lost*: adopt a growth mindset and practice "vicarious resilience" (see below).
- *If you are feeling the "fight or flight" physical sensations in your body*: try spending time alone to focus and breathe or utilize a guided meditation to calm your energy.

Utilizing the practices of reframing and mindfulness allows us to see that we are the student today growing from these opportunities. We then can be the mentors/teachers tomorrow to our professional colleagues and others we care about in our lives. These practices are fundamental toward building our resilience "muscles".

DEVELOPING AND FORTIFYING RESILIENCE

Resilience can be seen as the ability to adapt in positive ways to difficult and trying situations. It is learning to be flexible and to bend, not break, under challenging circumstances and to recover quickly from difficulty and perhaps continue to function relatively normally. When faced with adversity or significant sources of stress, these circumstances can be used to teach of us to be humble and to start again. Wisely, Nelson Mandela stated "The greatest glory in living lies not in never falling, but in rising every time we fall". He also believed that there are no "failures" in life, but rather "opportunities to learn".

Consider the five categories of resilience:

1. **Physical**—physical flexibility, endurance, strength.
2. **Emotional**—emotional range and flexibility, positive feelings, self-regulation, healthy relationships.
3. **Spiritual**—commitment to core values, flexibility, and tolerance of others' values, beliefs, and intuition.
4. **Mental**—increased attention span, mental flexibility, optimistic world view, incorporating multiple points of view.
5. **Social**—resilient people live in resilient communities.

The concept of "coherence" is where the resilience realms overlap with each other in a healthy center. In this space, we are in the "eye of the storm". It allows us to maintain perspective, to interact cooperatively, to better manage stress, to build networks in our community, to find and stay connected to our calling, and to live more authentically.

For resilience to be developed, we actually have to be subjected to stress in our lives. Interestingly, this is true not only for humans and animals but also for trees! The young tree must undergo the "stress" of being blown by winds in multiple directions to grow strong, flexible heartwood. This core of "resilient" heartwood then allows the tree to grow upright and to be strong enough to face storms in the future. I love that analogy and honestly have seen trying circumstances in my adult life in that way, as winds that are testing my ability to stay on my feet. I genuinely believe that the lessons that I learn from how I handled that challenge would inform me and make me more adept in the future "storms" of life.

> Try, try again! If you "fail", get up again!
> —**Angela Duckworth, Grit: The Power of Passion and Perseverance**[8]

"Resilience" is sometimes heard as a vague synonym for strength, or for vigor, but perhaps it is more helpful to consider resilience as the capacity to adapt and to continue forward in the face of struggle with your mental and physical health intact. Per Lucy Hone, adjunct senior fellow at University of Canterbury, New Zealand, and a director of the New Zealand Institute for Wellbeing,

> a huge component of resilience is mental and behavioral agility. One of the components of resilience is self-efficacy: the belief that you can navigate whatever you are facing. It does not mean that it is easy or that you must have a "stiff upper lip"; rather it involves all emotions and can come in fits/starts. Demonstrate self-compassion and quiet that critical inner voice and believe in your abilities to get through tough or distressing circumstances. Recall other times where you managed to take a deep breath and told yourself "I am doing the best I can. I can do this". And you did.[9]

Facing difficulties is what helps us develop what is called "self-efficacy belief". This is the confidence that we can meet the challenges ahead, just as we have met the ones that we have experienced in our past, and that we may still be experiencing. "Vicarious resilience" is where we draw upon those past experiences to inform us in the current challenge. We may not know exactly the skills/knowledge to bear, but we can utilize what we know and then courageously, humbly take one additional step at a time.

When the going gets tough, having social support is really important to bolstering you and providing perspective. Having financial security helps, having education helps, having more life wisdom helps, but none of these by themselves are the "silver bullet" to building resiliency. Every situation is different and different strategies work for different people. Sheryl Sandberg says in her excellent book, *Option B*, "We're not born with a fixed amount of resilience. It's a muscle we can build".[10] Like any workout, you have to be the one to do the work, but it can help immensely to have the right tools, knowledge, and a community that supports you in developing safe, healthy habits.

It is important to note that individual efforts to build resilience do not negate the fact that there are still huge external factors, such as social inequalities, structural racism, or underfunded support services, that individuals must navigate. There needs to be concurrent work on improving the overall health and equity of systems to make them a supportive environment for all. Perhaps, then, individuals would not then need to exert so much energy on building and maintaining individual resilience to function well in those spaces.

"Is what I am doing helping or harming me?" It is really important to know that you have a choice over the input into your world. Those that struggle to develop resiliency tend to have a fixed mindset as opposed to having a realistic appraisal of the situation and accepting "what is" without judgment. This mindful, less judgmental mindset helps to clarify expectations of yourself and of others. When we take the time to do this, we mitigate frustration and disappointment and even learn to be alright with mourning as we choose to let go of what we thought was going to happen. Clarifying intentions and goals that are value-based with less attachment to the exact details of the "how" and the "what" as well as the timeline takes practice and patience. However, this approach allows for increased creativity and less emotional angst overall.

Facing your fears can be liberating. It allows you to take control back and form an action plan when you ask: "what is the worst thing that can happen?" When you work through the "what if?" scenarios, you are then able to clarify what really is at the heart of a particular fear or anxiety about a situation. Each step allows you to ask, "is that true?" and "what would I do if that actually did happen?" Often at the end of this exercise, anxiety and other distressing emotions have diminished as you have tapped into your prefrontal cortex, and strategy-building has provided you a sense of perspective.

> *Only I will remain.*
> *I will face my fear.*
> *I will permit it to pass over me and through me.*
> *And when it has gone past, I will turn the inner eye to see its path.*
> *Where the fear has gone there will be nothing.*
>
> **—Frank Herbert, Dune**[11]

There is ongoing research that supports new thinking around resilience associated with loss and trauma. Professor George A. Bonanno is a professor of clinical psychology at Teachers College, Columbia University. His new book, *End of Trauma: How the New Science of Resilience is Changing How We Think About PTSD* (release date September 2021), points to the fact that most humans innately have a high capacity for resilience. This means that there is the ability of individuals exposed to highly disruptive events or to loss in their lives to maintain both healthy psychological and physical functioning as well as the capacity for positive emotions. In a 2020 interview, he discussed that it may be more appropriate to teach people to be flexible rather than resilient which involves "teaching people how to actively deal with stressors, so that they can take advantage of whatever resources, whatever traits, whatever strengths they have". The individual is then better able to figure out what is the best thing to do right here, in this moment.[12]

SISU

An important component of building your overall resilience is being compassionately humble with yourself and knowing that you do not have to have everything "figured out". When tough life circumstances show up, you may temporarily suffer from distress or uncomfortable emotions. You may, or may not, stay calm and collected. How you uniquely handle a particular situation is determined by your life circumstances to date, the cultural environment that you were raised within, and your current emotional/physical state of being. An example of this in my own experience is how I would handle a complicated emergent case at the beginning of my 14-hour shift versus at 4 am after a busy night on the overnight shift. I had more energy and overall bandwidth at the beginning of that shift, and it gave me the ability to have better perspective, emotional regulation, and cognitive processes.

I learned over time that when challenged with depleted personal stores of energy, I needed to slow down, be compassionate with myself, and not only double-check everything that I was doing but absolutely utilize the power of collective knowledge and observation of my caregiving teammates. This is called "resourcing". This is exactly the same concept as building your coping toolbox. Regular evaluation and fortification of both your internal resources (experience, skills, ability to use breath for emotional regulation and mindset refocus, etc.) and your external resources (supportive personal and professional community, online and hard copy knowledge resources, mentors, therapists, life-coaches, etc.) is quite honestly skillful and wise for a human navigating an ever-changing, often-challenging world.

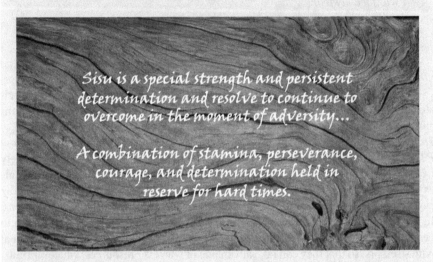

Sisu is a special strength and persistent determination and resolve to continue to overcome in the moment of adversity...

A combination of stamina, perseverance, courage, and determination held in reserve for hard times.

Opportunity: Consider these 10 ways to build resiliency by fortifying your resources:

- Make connections. Having good relationships; asking for and accepting help and support.
- Avoid seeing crises as impossible problems. How you perceive a situation makes it more of an opportunity than an insolvable problem.
- Accept change. Life is about change moment-to-moment whether we like it or not. Know that you have the ability to adjust your plans to reach your goals even when the timing or the exact path to that goal may need to evolve.
- Move toward your goals. Make your goals realistic and achievable (think of every New Year's blog you have ever read about creating goals!). It is more important to take small, actionable steps toward the larger goal than being immobilized by the perceived enormity and difficulty of reaching the said goal.
- Take decisive action. The antidote to anxiety and to procrastination is action! Move forward with clear intention and be patient with yourself.
- Look for opportunities for self-discovery. It is through hardship that personal growth occurs. Give yourself credit and appreciate your capabilities as you go!

- Maintain a positive self-concept. Build up your confidence that you can tackle difficulties that life presents to you. Learn to trust your inner wisdom and instincts. Remember these moments when challenges arise so that you can recall that you have successfully navigated trying circumstances before and survived. Are there tools/knowledge from those other instances that you can apply here?
- Keep things in perspective. Look at the broader context that the difficulty/challenge is arising within and note any tendencies that you may have to add story or judgment (of yourself or others) that may exacerbate the situation.
- Maintain a hopeful outlook. This is different than having "rose-colored glasses" on when perceiving a situation. When you look through the lens of optimism, rather than through fear, you will be better able to remain objective and move toward positive outcomes.
- Take care of yourself. Increased self-awareness and self-care are fundamental aspects of building resilience. You need to be in a healthy state of mind and body in order to have the capacity to be "nimble" and to intentionally act and speak when challenging circumstances present themselves to you.

In order to be our most resilient, healthy selves, we also need to create the time and space to practice all of these skills of self-care. Declaring ourselves worthy of care and drawing boundaries around time for ourselves is the next practice to explore.

BOUNDARIES—AN ESSENTIAL ELEMENT IN SELF-CARE AND REPLENISHMENT

Healthy boundaries are hard to draw and even harder to maintain. Maintaining good boundaries between personal and professional life helps to minimize the likelihood of work issues blending into your personal and family time. You need the time away from work to replenish your energy overall. Also, having the opportunity to meaningfully connect with others outside of work helps to decrease compassion fatigue and burnout from our challenging work environments. It is valuable to consider how we transition from work to our personal lives and vice versa. What practices do you currently have in place to make that transition clearer and more effective for you? An example may be debriefing critical incidents or emotionally challenging situations with our work teammates as a way of supporting each other overall, individually emotionally regulating, and decreasing empathic distress for everyone involved. Having the opportunity to process and integrate in a timely way means that these incidents and their "residue" are less likely to be in our bodies and minds when we return home to those that we love and care for.

When declaring a boundary around what you may need for yourself, be clear with yourself and with others on what that looks like. Consider what these guidelines, limits, and rules for ourselves look like. Equally it is important is to give some thought to how we will respond if someone steps outside of these limits or pushes them. These are not necessarily "rules" for someone else to follow, but they do need to be aware of the "rules" if they are to honor them. It is imperative that you are clear on what your boundaries are and implicitly and explicitly tell the world how you expect to be treated. Healthy boundaries are not harmful to others nor are they by nature selfish, mean, or permanent. Boundaries are about personal accountability and are intended to be protective of you and of your energy/time/resources. Be prepared to assert and to adjust boundaries as you and your life change.

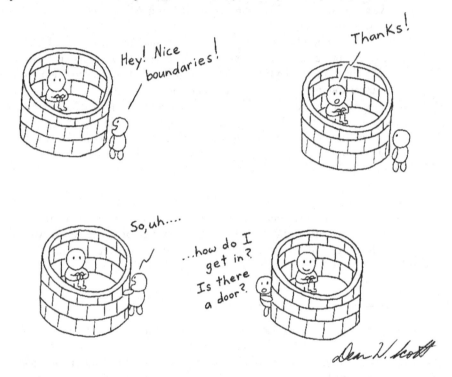

Healthy Boundary Creation—the "What" and the "How"

I found it helpful myself to have some further clarification about boundaries overall when I found myself evaluating where I have been willing and not to draw boundaries in my life to date. Here are some aspects that may be useful to you as well, as you ponder where you are in your boundary-making and holding skills:

Your bucket capacity—factors that may impact your personal boundaries:

1. Time. What are your time commitments and obligations? What can you say "no" to, so you can create time to say "yes" to something that you need or that restores you?
2. Content. What are you creating a boundary around (e.g., feelings, finances, privacy)?

3. Physical space. Feeling supported and safe.
4. Rights and responsibilities. Declare what you want and need; autonomy; respecting your beliefs and values.

Setting boundaries is essential if we want to be both physically and emotionally healthy. Creating healthy boundaries is empowering. By recognizing the need to set and enforce limits, you protect your self-esteem, maintain self-respect, and enjoy healthy relationships. In the setting of boundaries, it may be useful to consider what type of boundary-setting type category might you fall into? There are three types that have been defined: rigid, porous, and healthy. Our ability to create and enforce boundaries can be influenced by many factors: fear of rejection or confrontation, guilt, safety concerns, or perhaps we were not taught healthy boundaries. All of these, as well as other factors unique to you, your culture, and your upbringing, can inform your boundary-setting style.

These are some of the characteristics within each of the three types:

Common Traits of Rigid, Porous, and Healthy Boundaries

Rigid	Porous	Healthy
Avoids intimacy and close relationships	Overshares personal information	Values own opinions
Unlikely to ask for help	Difficulty saying "no" to the requests of others	Does not compromise values for others
Has few close relationships	Overinvolved with others' problems	Shares personal information in an appropriate way (does not overshare)
Very protective of personal information	Dependent on the opinions of others	Knows personal needs and is able to communicate them
May seem detached	Accepting of abuse or disrespect	Accepting when others say "no" to them
Keeps others at a distance to avoid the possibility of rejection	Fears rejection if they do not comply with others	

When establishing a boundary that is appropriate for you, you are not responsible for another person's reaction to the boundary you are setting. You are only responsible for communicating your boundary in a respectful manner. If it upset them, know it is their problem. Some people, especially those accustomed to controlling, abusing, or manipulating you, might push these boundaries. You can prepare for it by practicing what may be said, plan on how you will respond, and prepare to stand firm. Remember that if you allow a boundary to be broken or moved, it sends the message that it was not a boundary or that you can be swayed to do what someone else wants from you. You cannot successfully establish a clear boundary if you send mixed messages by apologizing. At first, you will probably feel selfish, guilty, or embarrassed when you set a boundary. Do it anyway, and remind yourself you have a right to self-care.

Setting boundaries takes practice and determination. Anxiety, fear, and guilt may arise but please do not let these impede you from taking care of yourself. When you feel anger or resentment or find yourself whining or complaining, you probably need to set a boundary. Listen to yourself, determine what you need to do or say, then communicate assertively. Learning to set healthy boundaries takes time. It is a process. Set them in your own time frame, not when someone else tells you.

Conflict Styles—Based on the Thomas Killman Conflict Modes Model[13]

I would like to share some additional concepts to better understand boundary-making and how to communicate those boundaries to others. When a conflict arises around a boundary that you are setting, which is really a goal for self-care, there are different approaches to the communication based on the relationship you have with the individual. The Thomas-Kilmann Conflict Mode Instrument has been used in a wide variety of capacities to better understand and then develop our individual and our team approaches to managing conflict. To navigate your approach more skillfully in developing assertive communication, consider the five styles of the Thomas Killman Conflict Modes Model and how they may play out depending on how high of a priority the goal (your self-care boundary) and the how important the relationship is to you:

1. Competing.
2. Collaborating.
3. Compromising.
4. Avoiding.
5. Accommodating.

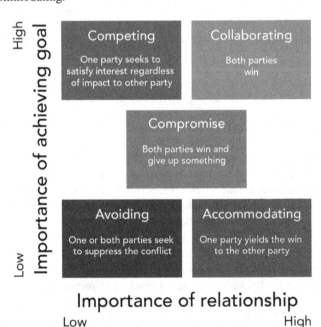

- Most individuals have 1–2 styles that they rely upon and use them when appropriate in situations:
  - Which to use and why?
  - How important are the issues to you? To the other?
  - How important is it to maintain a positive relationship?
  - Is the issue time sensitive (e.g., a dog that has been hit by a car and treatment needs to be determined immediately and quickly)?
  - To what extent does one party trust the other?
  - What is the power differential in the conflictual relationship?
- Escalator versus De-escalator language
  - **Escalator:** commands/certainties; comparisons; exaggerations; "other poisons" = shaming, ignoring, name-calling, anger, threatening, contempt, blaming
  - **De-Escalator:** "I" messages; uncertainties/open-mindedness; affirmations and acknowledgments; open-ended questions (get curious: "how?", "why?")

How Do You Know if You Are an Effective Boundary Setter with Another?
- You are responsible to each other, not FOR each other.
- You are receptive to and respect the other's boundaries.
- You can say "no" without guilt or feeling the need to apologize.
- You skillfully use boundaries to solve and to prevent problems or future conflict.
- You set boundaries in a high-brain state when possible (responding rather than reacting to decrease the emotionally charged nature of decision-making).

Examples to Practice Saying when Asserting Your Boundaries
- "I am not comfortable with this".
- "This does not work for me".
- "Not at this time".
- "This is not acceptable".
- "I can not do that for you".
- "Let me think about it and get back to you tomorrow" (gives you time to sleep on it!).
- "No". (This can be a complete sentence. You do not have to explain yourself. You have the right to determine what you do and do not want to do.)

How to Set Personal and Emotional Boundaries
- Define (identify) the desired boundary.
- Communicate (Say what you need).
- Stay simple (do not overexplain).
- Set consequences (say why it is important to you—not a justification but clarifying your values to yourself and the other).

Knowing Better = Doing Better
- Knowledge of healthy relationships (trust, respect, honesty, growth, ability to be alone).
- Healthy sense of self-identity.
- Self-esteem from within.

Practice:

- Using confident body language, face the person, make eye contact, use a steady tone of voice and an appropriate volume.
- Be respectful. Avoid yelling, using put-downs, or passive-aggressive behavior.
- Plan ahead. If possible, think about what you will say and how you will say it before entering a difficult conversation.
- Compromise when appropriate. Listen and consider the needs of the other person. You never *have* to compromise, but all healthy relationships need to have some give-and-take.
- Weigh the "yes:stress" ratio when you consider what you are willing and able to take on vs. not at this time (or at all).

SETTING BOUNDARIES AS PART OF WORK–LIFE INTEGRATION = HEALTHIER YOU

Four steps to wellness to protect yourself from compassion fatigue and vicarious traumatization:

1. Take stock of your stressors.
2. Look for ways to enhance your self-care and work–life integration processes.
3. Develop resiliency skills.
4. Make a commitment to implement changes.

Work–Life Balance vs. Integration

The concept of having work–life "balance" is honestly unrealistic. Rather than striving and causing ourselves additional stress as we aim for an exact balance of all the things that we value in both our work and personal lives, it is healthier to shift to the concept of integration. This takes all the moving pieces that make our messy lives wonderful and a struggle into consideration. We exist in an ever-shifting landscape of time and priorities which, when we apply self-compassion, can successfully result in a more fulfilling "work–life integration".

As a foundation of good self-care, good work–life integration is key. For those in the caregiving professions, this may seem like a particularly elusive and unrealistic concept. As stated by author Francoise Mathieu

... many helpers have very hectic shifts with few or no breaks and lead harried personal lives with numerous family commitments, errands, and chores and little time for rest and leisure activities. Helpers tend to be "on" from dawn to dusk ... some of us end up married to the job with no personal lives, taking on extra work shifts and collapsing for hours in front of the television when we get home.[14]

In order for successful integration of work and our lives to occur, there needs to be both personal accountability as well as a commitment from our organizations and leadership to supporting the wellbeing of the individual workers. For the individual associate, in order to bring a healthier allocation of energy/time to the different aspects of life, we might look for opportunities to bring a greater harmony into the areas that are of importance in our lives. Equally important, we must also give ourselves permission for refueling and time for necessary replenishment outside of work.

Concurrently, there has to be a larger conversation occurring for the organization and for teams to consider what schedule flexibility could be created to accommodate different lifestyles and needs of their associates. For some, working remotely on occasion may be an option. For others, it may mean looking at whether your shift start and end times could vary to accommodate personal and family needs. Studies have shown that workers want not more pay, but to have "greater control over work hours and more respect".[15] Another great paper out of the Headington Institute written by Laurie Anne Pearlman and Linda McKay titled "Vicarious Trauma: What Can Organizations and Managers Do?"[16] shared the following points around evaluation of how schedules might be adapted to better support individual caregivers:

- Flexible hours.
- Work from home where possible and appropriate.
- Permit those returning from medical leave to gradually build up to a full-time schedule.
- Train managers on how to support work–life integration.
- Encourage staff to stay at home with sick children or elderly relatives when needed.
- Eliminate unnecessary meetings (and/or allow for virtual participation).
- Communicate expectations clearly to staff.
- Allow for some autonomy in controlling their life priorities (as much as is possible).

Although the larger systemic conversations around creating a healthier workspace and lifestyle for veterinary professionals is occurring, the onus is upon each of us as individuals to control and seek to improve what we can: our own mindset and unique approach to wellbeing. It may also be appropriate to consider having clarity through self-reflection on what you need to have documented in your work contract, particularly around schedule expectations.

Dr Cheryl Richardson is a well-known life coach, speaker, and author of books on self-care, finding balance, and assertiveness. One of her best-sellers is her book, *Take Time for Your Life* (2009). Her writings as well as many others, such as Arianna Huffington's book, *Thrive* (2014), and website, ThriveGlobal.com, have contributed to my thinking on the following initial suggestions for each of us to consider putting into action.

Practice:

- Explore reasons for roadblocks that prevent you from "getting the life that you want".
- Take time to reflect upon "What fuels you? What drains you?" (Write these responses down!)
- Need to create realistic (read: achievable) goals.
- Create a supportive community (accountability partners and allies in positive change).
- Give yourself permission, without guilt, for self-care time:
  - Exercise/physical activity.
  - Fun, restorative activity.
  - Down time (e.g., unstructured time).
- Scheduling is necessary (use a planner and be proactive).
- Collect ideas from others! Crowd source on creative ideas but also on strategies of implementation.
- Take stock of your stressors and your coping skills. Make a commitment to implement changes that will serve you and your personal/professional goals.

Why should you invest the time and energy in this process?

1. Fortifying your resilience for challenging circumstances that arise in life.
2. Longevity within our chosen career paths (holistic wellness practices to decrease compassion fatigue and vicarious traumatization).
3. Improved overall wellbeing (physical, mental, emotional).

Creating the Plan That Is Best for You!

There are five necessary components to developing an intentional and actionable plan: *awareness, assessment, plan, accountability partner, and follow-up.* Consider using the proven *SMART* technique to develop your self-care goal to develop this wellbeing plan:

S = Specific. What exactly do you want to accomplish?

M = Measurable. How will you demonstrate or evaluate that your goal has been met?

A = Achievable. Is the goal reasonable? (Make the step realistic and small enough to achieve = a *micro-step* on the larger path toward your self-care goal.)

R = Relevant. Is the goal aligned with your overall objectives, intentions, and values?

T = Timely. Is there a deadline or other way that you can measure whether your goal is successful?

Other *Idea*s to *Get Your Creative Cogs Turning*

1) **Letting go of work during off-hours**
   - Embracing activities that are fun, stimulating, inspiring, and generate joy.
2) **Acquiring adequate rest and relaxation**
   - Concept of prioritizing rejuvenation and replenishment (what depletes you and what fuels you?).
   - Tailored to your own interests and abilities.
3) **Practicing effective daily stress reduction methods**
   - Tailored to your own interests and abilities.
   - Manage stress during work hours and off-hours.
4) Making time for **social connection and support** with family, friends, and professional colleagues.
5) **Collaborative leadership**—when to rely upon yourself and when to delegate.

We have examined the self-kindness and strength in community components of the self-compassion principle. The third and really impactful "ingredient" is mindfulness. There are entire books and courses dedicated to the concept of mindfulness. Importantly, there are Western teachers, such as Thich Nhat Hahn, that have taken the fundamental philosophies and practices from Eastern Buddhism and created secularized approaches to the core values and teachings. I have included some resources for you in the Recommended Readings but here I wish to touch on how mindfulness supports our self-care practices.

MINDFULNESS AS A "SUPERPOWER"

> Between stimulus and response there is a space. In that space is our power to choose our response. In our response lies our growth and our freedom.[17]

> *—Viktor Frankl, Austrian neurologist, psychiatrist,*
> *philosopher and Holocaust surviror.*

Mindful awareness allows us to move from the reactionary parts of our brain, such as the amygdala, into our "higher" mind, which is the forebrain and anterior cingulate nucleus. This awareness allows you to be the objective, curious observer of the actual circumstances as they are occurring in that moment in your life. Our perception of what "is" can be influenced by fear, bias, habits, and/or wishful thinking. When we intentionally choose to be more mindful in our interaction with the present moment, it refines and clarifies our attention. This allows us to respond with purpose and integrity when we are more fully present. This is the space that the famous author and psychologist, Dr Frankl was pointing to in the quote in which growth and freedom to choose to occur. Stephen R. Covey, well-known inspirational author, referred to both Dr Frankl and to the power and freedom in mindful attention as well in his widely read book "The Seven Habits of Highly Effective People" (1989).

Thinking vs. Awareness

The state of "calm" historically did not come easily to me. I have a lot of energy and a tendency toward constant motion. When in a veterinary practice environment, this energy and movement was also imbued with a sense of urgency. This could, and did, translate into agitation, activating the entire cascade of thoughts, emotions, and physiologic reactions that are likely to be triggered in the sympathetic nervous system response. What I was not aware of at the time was how much this was contributing to a vicious cycle of reactive feelings and body sensations for myself and my teammates.

As humans, we have this amazing built-in capacity for "emotional attunement". This was part of our evolutionary "operating system" as social pack animals that required a community for safety and survival. This subconscious, physiologic capacity combined with "mirror neurons" that were discovered in our brains in 1990s research, allows us to literally feel and then reflect what others are feeling. Take a moment to consider this for yourself: how much are your emotions, your facial expressions, the tone of your voice impacting not only the way that you are feeling in the moment but also directly impacting others (humans and animals) around you?

I was fortunate enough to have mentors and colleagues along the way that provided me positive role models of what "calm in the midst of the storm" can look like in both my personal and professional life. I also became increasingly aware of when individuals behaved or spoke in ways that felt in conflict with my values and self-identity. With all of that in mind, can you perhaps then, recognize how impactful it is for you and for all those around you to bring mindful attention to your words and actions? Notice when you get sucked into the whirlwind of agitated or righteous thinking. When unhelpful, distressing emotions arise, they can become entangled with those unhelpful thought patterns and create a cycle of unproductive rumination. What if instead, you "slowed your roll" enough to notice when you are feeling balanced, clear-headed, and positive. What is going on in that moment that you can tap into again as a resource in the future when you become agitated? This self-awareness decreases distress for you and it can also decrease the potential for unintended harm you may cause to others around you when you react rather than respond intentionally.

Impacts of Practicing Mindfulness

There has been so much research done over the last 30 years into the how, what, and impacts of secular mindfulness practices. There are now known and verified impacts of mindfulness on many wellbeing realms: it calms the body, decreases stress, reduces anxiety, integrates emotions, and relieves chronic pain.

Some of the physiologic pathways that are involved in these positive impacts are already known, and some are yet to be discovered. Tapping into the innate operating system that our human bodies come equipped with for better health and equanimity with our communities is something that all of us have the power to do, but we have to have an awareness of the how and of the what.

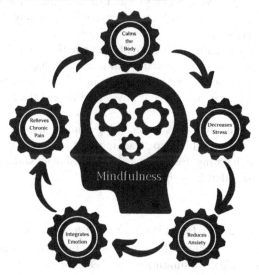

Practices That Support Development of Our "Mindfulness Muscles"

Mindful Awareness Practice

We know that we have an unconscious judgment of situations. Humans are drawn to narratives as moths are to flames. The narrative frames the experience and helps us to make sense of situations. Our brains are wired to attach meaning and to create these stories. It also shifts our perspective from "experiential self" into "narrative self". This can help to provide both a sense of meaning and fortify our self-identify in the world. It takes patience and practice to notice when we move from our actual experiencing self into the sense of self that we have created in our heads.

If our awareness is the empty stage, thinking is the cast and screenplay. Zoom in and zoom out. Indulging every thought is a habit we carry throughout our lives. Starting with your mindset and your attitude, can you shift it to a good-natured, affable mindset, one of curiosity toward this new experience of mindful awareness?

Thinking is often an unconscious process, simply part of our existence. We can have upward of 12,000 thoughts a day, and 75 percent of them are subconscious! Our minds secrete thoughts the way that our stomachs secrete digestive enzymes. Neither good nor bad, just a part of being a living human. Shifting our perspective toward one of awareness of when we are thinking can provide a sense of freedom and of more available time and energy as you reclaim your attention. This is a valuable process, as we all have a sense of time and energy scarcity. Perhaps noting what type of thinking process is occurring can also be useful to slow you down, to notice what thoughts are arising and see if you can catch patterns of thinking/feeling that show up more regularly (you can even title them "Story #1", "Story #2", and such). Noting of thoughts may be a quiet inner repetition of "analyzing, analyzing", "planning, planning", and "worrying, worrying".

Can you practice letting both thoughts and emotions roll through you like a weather system? Consider instead, what is actually happening in this very moment. Where are you? What are you sensing? How is your breathing? Can you increase your awareness of your physical body and the environment that you are in?

Please note: If you find yourself becoming agitated, look around to get perspective, calm yourself with deep steadying breaths, and stop the practice if it becomes overwhelming. Be gentle and compassionate with yourself—keep trying but at a pace that feels right for you.

Mindful Attention—Energy Flows Where Your Attention Goes

- How you respond to what is happening is the focus of your attention.
- Let go of self-defeating, limiting habits, behaviors, and thinking patterns.

- Accept the present moment for what it is—lose the story, the judgment, the knee-jerk reflex of your reaction.
- Create a more vibrant and alert space for you to be in your integrity; be in alignment with your values and authentic being by practicing the shift from reactionary amygdala and hippocampus into the forebrain, where thoughtful, creative responses occur.
- Small practices along the way to build your mindful muscles—be patient and give yourself a plenty of grace as you figure out what works for you.

Gratitude to Change Your Attitude

The amount of research in the healing and health-fortifying capacity of practicing gratitude is great. There are many excellent articles, lectures, and books that have been dedicated to the myriad of ways that gratitude practice positively impacts feelings of happiness, purpose, and connection. The power of practicing gratitude is literally transformative in impacting mindset as it rewires our brains! Research efforts using fMRI revealed that there was greater neural sensitivity in the medial prefrontal cortex, the brain area associated with learning and decision-making, in people regularly utilizing gratitude practices (journaling, "paying it forward", reframing using positive words, emotions, and attitudes; and developing optimism).

Studies have also demonstrated that those who practice both mindfulness and gratitude on a regular basis had less emotional reactivity to potentially stress-inducing circumstances, as their brains shifted more to the prefrontal cortex and less to the hippocampal and amygdaloid regions, which are associated with distressful emotions. We know that, in adult humans, our thinking and perspective is often through the filter of our preprogrammed negativity bias. This is part of our evolutionary survival instinct that exists to heighten our awareness to possible threats and guide us toward survival. Neuroplasticity is now known to allow us to practice reprogramming ourselves to think more positively, shift toward optimism, and have faith in a better future.

EXAMPLES OF SOME GRATITUDE PRACTICES:

- Keep a journaling or use your mobile device to take notes (there are so many apps to support this practice) on what is going well for you today—write down and elaborate on 3–5 things (or more). Notice the calm and the centering that may occur as you focus on the good that you have in your life to counterbalance the more challenging aspects that also exist.
- Use mindful meditation practices which can be as simple as reflecting upon three things that you are grateful for in this moment, as you also take time to deeply breathe or go for a walk outdoors to broaden your perspective (lift your eyes to the horizon; "nature pill").
- Write to someone who is important to you or who has inspired you (whether you send the letter or e-mail or not!). It is the practice of being in the space of gratitude and feeling cared for that can release oxytocin

(one of the "feel good" hormones that connects us to our human and furry family members).

- Practice reframing—when faced with something that you feel is a negative or is really challenging, you can ask yourself: "What are the lessons to be learned here? Where is the good in this situation?"
- Use gratitude triggers, placing objects that in your home or workplace that trigger gratitude when you see them, e.g., a favorite vacation memento, a picture of your pet(s) or child(ren), an inspirational phrase that speaks to you.

Mindful Micro-breaks—"Mini Resets" of Your Brain and Nervous System
Sit. Stay. Heal.

> —*Thich Nhat Hahn*, **Zen Master and pioneer**
> **in modern mindful practices**

Time is an interesting thing. It is a human-created construct that we have allowed to significantly impact our entire existence. So much of the world has bought into the idea of time scarcity—that there are not enough hours in the day to get everything done. Sound familiar? The pressure to "do" and to "multi-task", particularly in Western culture, can make our lives feel out of control and that we do not have an extra moment to spare, particularly for self-care.

I get it. I, like so many, wore my "busyness" as a badge of honor. In retrospect, I also now can see that busyness contributed to my sense of worth in the world. It effectively would distract me from slowing down, making it easier to ignore uncomfortable thoughts and emotions. You only get one crack at this life, and as Ferris Bueller said, "Life moves pretty fast. If you do not slow down, you may miss it". There are proven techniques to not only be more productive but also to give you a genuine feeling of time slowing down enough for you to be more present in that moment. Mindfulness practice includes intentionally taking mindful "mini-resets" throughout your—it is a proven method.

In his New York Times article, Tim Kreider describes taking breaks this way: "taking a break is not just a vacation, an indulgence, or a vice; it is as indispensable to the brain as vitamin D is to the body, and deprived of it, we suffer a mental affliction as disfiguring as rickets … It is, paradoxically, necessary to getting any work done".[18] Research over the last several decades with human healthcare givers, particularly nurses in busy clinical environments, has demonstrated that taking even eight minutes every few hours can restore energy, increase focus, and decrease the likelihood of medical errors. These regular mini-resets also help us to be more resilient when stressors arise, and they can function as an opportunity to help us cope in a healthier way with our daily grind. I love that in Japan, they have a nap break concept known as "inemuri", which is literally translated as "sleeping when present". This is a means to boost energy and increase cognitive processing. The Mayo Clinic has a fair bit of research on the effects of long shifts and emotionally, physically challenging work environments. One study on meditation practices found that investing a few minutes per day (10–15 minutes seemed to be the sweet spot) can ease stress and help decrease depression, fatigue, high blood pressure, and insomnia.[19]

So, to summarize why these micro-breaks are valuable to consider incorporating into our daily lives:

| INCREASE FOCUS | INCREASE PRODUCTIVITY | IMPROVE VITALITY AND ENERGY LEVEL | DECREASE BLOOD PRESSURE AND ANXIETY | IMPROVE OUTLOOK AND ABILITY TO CONNECT |

"Micro-steps" = too-small-to-fail, science-backed strategies for real behavior change

- Setting an intention at the beginning of your day
- Setting a self-care alarm for a break that is at least 10 minutes long and is non-work related
- Declaring an end to your workday

Impacts of Micro-Breaks

1. Movement breaks are essential for your physical and emotional health.
2. Breaks can prevent "decision fatigue" and boost morale.
3. Breaks restore motivation, especially for long-term goals. According to author Nir Eyal, "When we work, our prefrontal cortex makes every effort to help us execute our goals. But for a challenging task that requires our sustained attention, research shows briefly taking our minds off the goal can renew and strengthen motivation later on[20]".
4. Breaks increase productivity and creativity. Working for long stretches without breaks leads to stress and exhaustion. Taking breaks refreshes the mind, replenishes your mental resources, and helps you to be more creative.

MINDFULNESS AND SELF-COMPASSION

Mindfulness is loving awareness of *moment-to-moment experience.*
Self-compassion loving awareness of the *experiencer.*

Mindfulness asks, "What do I *know*?"
Self-compassion asks, "What do I *need*?"

Mindfulness regulates emotion through attention regulation
Self-compassion regulates emotion through affiliation

Mindfulness is calming
Self-compassion is warming

(https://mindfulnessexercises.com and Sean Fargo)

Words of wisdom from Dr Lynn Roy, small animal practitioner, student wellness advisor at the Cummings School of Veterinary Medicine at Tufts University:

When I graduated vet school 36 years ago, wellness or wellbeing were not words yet, and when you chose veterinary medicine as a profession, you were expected to accept that this was to be the primary focus of your life. Work/life balance also was not "a thing", as your work was your life. Women in veterinary medicine were still a minority at that time, so I always felt like I had to work even harder to "prove myself" to every employer and colleague. Working 7 days a week, on call every other night, and no time off until after you worked there for a full year was the norm. This was also a time when local veterinarians did most of the surgeries, work ups, and emergency coverage without the benefit of specialists, advanced imaging, and emergency care facilities. But clients trusted their veterinarian, and lawsuits were not yet a common occurrence which made for strong relationships with clients. The human–animal bond was evident but not as strong as we see today, as most cats and dogs spent most of their time outdoors and not in the owner's bed. The absence of the internet was a help and a hindrance. It forced us to have veterinary medicine all in our head, as there was no "VIN" (Veterinary Information Network) to help us out, and the knowledge, wisdom, and clinical experience of older veterinarians was worth its weight in gold. However, there was not Facebook either to allow a negative experience or misinformation to be communicated at lightning speed. So, some stresses are the same and some are different … . But there will always be stresses in our profession, as we are always doing our best to "save the world"—or at least our little part of it! What I would say to a younger me is, find your voice … early. Tell your employer/manager when you are overwhelmed, need help, need time, and become a good self-advocate. This can all be done in a very polite and professional manner. It is important to be authentic and real to everyone: your clients, your colleagues, your friends, and family and take time to communicate. Communication is key to maintain every good relationship. Take care of yourself … . There is only one you, and you are worth it! If you are in a toxic environment and have communicated well and given it some time to improve, but there has been no change, it is ok to leave. Don't "burn any bridges", but you can look for another job or even [a] career path that better suits your skills, time, and needs at that point in your life. Remember our veterinary journey is a marathon, not a sprint; take the time you need to grow and develop not only as a veterinarian but as a person. Take care of the whole you!

NOTES

1. Figley, CR. (2017), *The Green Cross Academy of Traumatology Course Workbook for Compassion Fatigue Educator Certification* (https://www.figleyinstitute.com).
2. Firth-Cozens, J. (2003). Doctors, Their Wellbeing, and Their Stress. *British Medical Journal*, 326 (7391), 670–1.
3. Fahrenkopf, AM, et al. (2008) Rates of Medication Errors among Depressed and Burnt Out Residents: Prospective Cohort Study. *British Veterinary Journal,* 336 (7642), 488–91.
4. Oxtoby, C, et al. (2015). We Need to Talk About Error: Causes and Types of Error in Veterinary Practice. *Veterinary Record,* 177 (17), 438–44.
5. Salvagioni, DAJ, et al. (2017). Physical, Psychological and Occupational Consequences of Job Burnout: A Systematic Review of Prospective Studies. *PLoS One,* 12, e0185781 (https://journals.plos.org/plosone/article?id=10.1371/journal/.pone.0185781).
6. Hanson, R. (2018). *Resilient: How to Grow an Unshakeable Core of Calm, Strength, and Happiness.* New York: Harmony.
7. Brach, T. (2016). "Real but Not True: Freeing Ourselves from Harmful Beliefs". Recorded presentation (https://www.tarabrach.com).
8. Duckworth, A. (2016) *Grit: The Power of Passion and Perseverance.* New York: Scribner.
9. Hone, L. (2020) The Three Secrets of Resilient People/TED Talk (https://www.ted.com/talks.lucy_hone_the_three_secrets_of_resilient_people).
10. Sandberg, S, and Grant, A. (2017). *Option B: Facing Adversity, Building Resilience, and Finding Joy.* New York: Knopf.
11. Herbert, F. (1965) *Dune.* New York: ACE.
12. Armstrong, K. (2020). "Remarkable Resiliency: George Bonanno on PTSD, Grief, and Depression". *Association for Psychological Science* (https://www.psychologicalscience.org/observer/bonanno).
13. Thomas, K, and Kilmann, R. (1975). The Social Desirability Variable in Organizational Research: An Alternative Explanation for Reported Findings. *Academy of Management Journal*, 18 (4), 741–52.
14. Mathieu, F. *The Compassion Fatigue Workbook.* New York: Routledge, 98.
15. Duxbury, L, et al. (2008) "Too Much To Do, and Not Enough Time". In K Korabik, et. al. (Eds.), *Handbook of Work-Family Integration: Research, Theory, and Best Practices* (pp. 125–40). London: Elsevier.
16. Pearlman, L, and McKay, L. (2008). "Vicarious Trauma: What Can Managers and Organizations Do?" (Excerpted from "Understanding and Addressing Vicarious Trauma") Headington Institute, New Zealand (https://ovc.ojp.gov).

17. Frankl, VE. (1946). *Man's Search for Meaning.* (Reprinted by Beacon Press, Boston in 2006).
18. Krieder, T. (2012) "The 'Busy' Trap". *The New York Times.* (The Opinion Pages, June 30, 2012).
19. Sood A. (2013). Relaxation, Meditation and Prayer. In: *The Mayo Clinic Guide to Stress-Free Living* (pp. 262–78). Cambridge, MA: Da Capo Press/Lifelong Books.
20. Eyal, N. (2019) *Indistractable: How to Control your Attention and Choose Your Life.* Dallas: BenBella Books, Inc.

5 Change Is in the Wind—A New Framework

In a gentle way, you can shake the world.
Your beliefs become your thoughts,
Your thoughts become your words,
Your words become your actions,
Your actions become your habits,
Your habits become your values,
Your values become your destiny
You must be the change you wish to see in the world.
 —Mahatma Ghandi quote/poem, Create a Sacred Space of Awakening

When we reframe what we are working toward and for, we are more likely to have positive outcomes for ourselves, as well as our patients. By connecting to our personal purpose, values, and community, we can flourish in our personal and professional lives.

I admit to being an optimist at heart. My natural inclination is to be positive and stay hopeful. This has been challenged so many times since I entered veterinary school and went into clinical practice. I want to be clear that I am not being naïve or stating that concentrated effort will not be needed to shift and modernize our professional culture toward one that supports all caregivers and meets our clients' and veterinary patients' needs. Decreasing stigma around mental health, increasing permission for self-compassion, and embracing that we can have a holistically healthy life while being rock star veterinary professionals? Yes, please!

I sincerely believe in the capacity of this amazing community of brilliant, selfless souls to lean into this work for every one of us to feel welcome, to thrive, and to stay connected to our professional purpose. The predominantly unhealthy professional mindset, culture, and work habits that exist in veterinary medicine worldwide are long-standing and deeply entrenched. However, if there is another thing I have seen in myself and other veterinary professionals, we are a determined group when we put our minds to it! We also collectively know that the time for uncomfortable conversations, new approaches, and evolution of our beloved profession has come.

The journey of this book thus far has been to take stock of what we have now come to understand about the historical culture of the veterinary profession, to examine the potential hazards for those in caregiving roles, and to raise our own

DOI: 10.1201/9780367816766-5

self-awareness and compassion. My intention was to frame the conversation such that we can be better informed and give ourselves permission to declare our individual and collective worth and ethical right to care as well. We deserve to feel valued, safe, and to experience less suffering. There are veterinary profession-specific issues, e.g., high debt/lower income for recent US veterinary graduates, that have led to increased anxiety and decreased work satisfaction. There are also many more work-related stressors that arise from dysfunctional systems and a culture that we have identified as negatively impacting both human and veterinary medicine. Do you want the good news? There have been robust efforts over the last several decades from around the world to allow us as caregiving communities to identify and better understand the nature of our occupational concerns. "Name it to tame it!" When we know what we are experiencing, and why, we can then strategically seek sustainable solutions to improve the quality of our lives.

It has only been in the last 15 years or so that there has been a distinct increase in the international research endeavors examining mental health in the healthcare professions, including veterinary medicine. Interestingly, it is only in the last 3–5 years that the audible "buzz" has grown in veterinary circles large and small about concern for our veterinary colleagues' wellbeing. These discussions have led to a closer examination of our classroom and workplace culture. There is renewed scrutiny of training curriculum and structure. There has been an uptick in research endeavors, wellbeing "committees" at practice and organization levels, and interest from the global veterinary to examine how best to implement the change that is clearly needed and being asked for by today's veterinary associates. Almost every veterinary association now has committees dedicated to wellbeing. WSAVA now has a dedicated committee, the Professional Wellness Group, that examines and supports education and resources for veterinary professionals worldwide.

Despite these clear indicators that there is heightened awareness and interest in seeing our professional colleagues flourish in their chosen veterinary field, skepticism and cynicism also persist. Most of us drawn to the veterinary profession would classify ourselves as scientists and "geeks" who value analyzing systems and research data. Interestingly, much of the existing literature on mental health in the veterinary profession tends to focus on "mental distress" which "emphasizes factors related to burnout, stress, anxiety, depression, and suicide".[1]

In Jean Wallace's 2019 paper that examined the positive aspects of veterinary work, she shared that her literature review revealed that there were about twice as many articles that discussed veterinary wellness that focused on the *negative* aspects of mental health (e.g., stress, burnout, suicide) compared to *positive* aspects of wellbeing.[2] I call this out as a nod to our natural tendency toward the negativity bias. I love the axiom that our human brains are "teflon to the positive and Velcro to the negative". We know this to be true, but are we truly aware of how much that impacts not only our daily lives but also the way that we approach the entire topic of wellbeing in the veterinary profession?

Herein lies the secret to our personal and professional community learning and growth opportunities: the time has come to reframe the entire conversation! I am excited to share with you in where there are current, and developing, conversations and initiatives around veterinary professional wellbeing awareness and training—but presented through a more positive lens. There are three important areas of research, from which I want to share highlights with you in this chapter, that are vital to counterbalancing any of the negative aspects of our caregiving work:

1. Compassion satisfaction.
2. Positive psychology and the concept of *flourishing*.
3. Eudaimonia.

Each of these has an important common foundation: staying tethered to your "why". One of the most fundamental tenets that makes all the effort worthwhile in our caregiving work is to "remain connected to our values and to our purpose". What do we each uniquely find invigorating and fulfilling in our work that keeps us engaged and in the veterinary profession? This examination is vital for us to do as individuals as well as a community to best assess where we are currently and what are our collective aspirations.

An entire chapter (or book!) could be dedicated to each topic. My intention here is to be sure that you are introduced to them to add a new lens to view all that we have discussed thus far in this book. Equally, I am optimistic that these may serve as a more helpful framework for future discussions and initiatives as we seek protective measures against compassion fatigue, empathic distress, and burnout.

COMPASSION SATISFACTION

Compassion fatigue has an important counterpoint: *compassion satisfaction*. Compassion satisfaction comes from the positive and fulfilling aspects of changing and saving lives. It is valuable to reflect upon the rewards of our work and where they come from. What sustains us and keeps us coming back to work shift after shift, even when the work is challenging, if not downright exhausting? What is your professional "why"?

> Before starting your workday, take a moment to literally stop in your tracks and ask yourself, "Why am I doing what I am doing?" After you hear your answer, remind yourself gently that you are making a choice to do this work. Take a deep breath; breathe in both the responsibility and the freedom in this acknowledgement.
>
> —*Laura Van Dernoot Lipsky, Trauma Stewardship*

Veterinary caregivers feel satisfaction when we know that we have made a positive difference with our patient's wellbeing, with pet owners, and within our larger profession. In contributing to the greater good and knowing that your care had a direct impact on the betterment of animal wellbeing, there is an increased sense of worth as a caregiver and as a person. In an Humane Society of the United States research effort in 2018, it was found that the top three stressors for veterinarians were: difficult/non-compliant clients, not enough time to get the work done, and discussing (disputing) financial matters with clients. On the other hand, the top three satisfiers for veterinarians were: helping and healing patients, grateful clients, and working as a collaborative team toward best patient care.[3] In a recent paper, Dr Vanessa Rohlf and her co-authors sought to identify predictors of professional quality of life for veterinary professionals. As would be expected, they found that if there is increased professional fulfillment and compassion satisfaction, individuals would be less likely to leave the workplace (or the profession perhaps?). Per the authors, "personal factors (optimism and reframing) and workplace factors (opportunities for development) predicted compassion satisfaction in vets and vet nurses".[4]

In the 2019 Merck Animal Health Veterinary Wellbeing Survey, the researchers dedicated an entire section of the study to "job satisfaction".[5] They examined 14 dimensions that examined the individual's awareness of whether their current job and the workplace/organization supported an associate feeling valued and fulfilled. There were also several questions about the practice culture/environment including team dynamics. Of the 2971 respondents surveyed, "being invested in their work" and "making a contribution" were the two aspects that were most valued. The top drivers of job satisfaction were:

1. Good work–life "balance" (Note: this was also determined to b the biggest differentiator as to whether someone felt distressed or not distressed and had a higher sense overall of wellbeing).
2. Enjoying the work.
3. Being paid fairly (survey results supported that respondents' satisfaction with various compensation systems, e.g., salary vs. salary +production, were similar).
4. Having a supportive relationship with coworkers.

In addition to the satisfaction that can be supported by a healthy work environment and work–life integration structure, the actual demonstration of compassion is beneficial. It is a two-way connection. The compassion that you provide to another, be it animal or human, activates your parasympathetic nervous system, and there is a soothing impact on you, the giver, of compassion. This is part of the "healing of the healer" and "helper's high" that we experience when we feel connected to our veterinary patients and to the humans around us. Connecting more, leaning in, being present, listening without judgment, and holding space for another's suffering are all ways that we may demonstrate compassion to another. Practicing of compassion, particularly when the circumstances are challenging and uncomfortable (for you, for the other) improves efficacy. Learning how to become more effective and skillful with compassion practices will make a significant difference to both you and the recipient.

Importantly, compassion satisfaction is also reinforced with healthy coping strategies (e.g., boundaries between work and personal life) and regular self-care. Only you can know what circumstances will support your feelings of happiness and of fulfillment. It is vitally important that we have this awareness and accountability around our individual wellbeing as we also collectively strive to develop healthier work environments that support us as caregivers. Compassion satisfaction is a unique aspect of the professional quality of life and can coexist alongside those same elements that contribute to compassion fatigue and empathic distress. The goal is to bring increased awareness and to fortify those practices and the mindset shifts that promote increased satisfaction.

> If you are interested in an evaluation of your own current level of compassion fatigue, compassion satisfaction, and possible degree of burnout, consider using the evidence-based, Professional Quality of Life Survey (ProQOL) developed by Beth Stamm and Charles Figley (http://www.proqol.org). This is a nondiagnostic, pulse point evaluation tool to gauge the current cost of fulfillment that is being experienced by a caregiver.

POSITIVE PSYCHOLOGY—SUPPORTING *FLOURISHING* IN OUR LIVES

Positive psychology is a branch of psychology focused on the character strengths and behaviors that allow individuals to build a life of meaning and purpose—to move beyond surviving to flourishing. Martin Seligman is regarded as the "founding father" of positive psychology who promoted the concept while he was president of the American Psychological Association in 1998. He, along with other colleagues that were proponents (Christopher Peterson and Mihaly Mealy), built upon the work of well-known humanistic psychologists like Abraham Maslow who first used the term "positive psychology" in the 1950s.

Those involved in research and practice in this field seek to identify and evaluate the elements of what constitutes a "good life". There is an emphasis on sustained meaning and deep satisfaction, not just on temporary states of "happiness" that are contextual in nature. The primary intention of positive psychology is to identify and build mental assets rather than the traditional construct of evaluating "weaknesses" and addressing "problems" in life as they arise. Identifying one's core virtues (such as courage and wisdom), character strengths (such as persistence, kindness, mercy, and curiosity), utilizing the power of positive emotions, and focusing on what makes your life meaningful are all components of applying the tenets and practices of positive psychology.

Dr Nadine Hamilton, a psychologist who lives in Queensland, Australia, has had a passion around supporting veterinary wellbeing for many years. Her doctoral research focused on stress, burnout, and suicide in the veterinary profession. Dr Hamilton has had several initiatives and platforms that showcase and share her efforts to apply positive psychology to the myriad of wellbeing concerns that unfortunately abound in our veterinary community and environments. In her excellent 2019 book, *Coping with Stress and Burnout as a Veterinarian*, Dr Hamilton shares her wisdom and guidance around many topics that directly apply to many veterinary caregivers' lives. On the topic of positive psychology specifically, she provides excellent descriptions of these concepts and how they may be utilized to create a fortified sense of self-efficacy and confidence around sustainable wellbeing.[6]

The PERMA model that takes the core values and character strengths into consideration is a multifaceted tool to consider adding to your wellbeing toolbox that fortifies overall flourishing in one's professional and personal life.

This is what is meant by the PERMA model:

P—Positive emotion (feeling good, experiencing positive emotions, such as optimism, hope, and flexibility about change).

E—Engagement (having the experience of fulfilling work, interests outside of work, and experiencing the concept of "flow", a state of concentration and effortless enjoyment).

R—Relationships (social connections, love, intimacy, emotional and physical interaction).

M—Meaning (having a purpose, finding meaning in life, knowing your "why").

A—Accomplishments (ambition, realistic goals, important achievements, pride in yourself).

To support the use of the PERMA framework for yourself, it may be helpful to utilize a free character strengths questionnaire to find out your individual character strengths on the website: http://positivepsychsolutions.pro.viasurvey.org.

I know of several other friends in the veterinary profession who have pursued certification in positive psychology. Josh Vaisman and Dr Phil Richmond were generous in sharing each of their perspectives on how positive psychology has impacted their own wellbeing and that of the veterinary teams introduced to the philosophy and practical applications through their respective training and coaching.

Josh has been involved in veterinary medicine in a variety of support roles since the late 1990s. He certainly had the opportunity to see both the great sides of practice and to experience the innate challenges associated with working in veterinary medicine. His professional journey led him to a place that he thought would be fulfilling where he was a business owner, in a leadership role, and was contributing to the profession with practice management consulting. Josh shared that despite his feeling successful professionally and having a personal life that supported him, he hit an emotional brick wall in late 2016.

The overwhelming physical and emotional exhaustion took him by surprise. He shared that he simply could not make himself go into work "one more time". In reflection, he came to see that there were many aspects of his work life that were not serving him. Within six months, he completely transitioned out of all of the prior work-related roles and gave himself permission to take several months to breathe, recoup, and to take stock. One of the important things that Josh realized was that there was a strong need for positive culture change at a systemic and organizational level in the veterinary practice world in order for veterinary associates to truly thrive and flourish. This led him to explore and then pursue a master's degree in applied positive psychology and coaching psychology. This is not about "thinking positive" or having some naïve "all will be well" perspective on the world. Rather, in Josh's words, positive psychology

> is about cultivating tools, practices, and mindsets that allow us to respond productively to and thrive despite the inevitable and normal challenges in life. At the organizational level (and in the veterinary profession), the science of positive psychology provides a toolbox for leaders to craft environments in which veterinary professionals have the maximum opportunity to thrive in their work and enjoy the sustainable fulfillment they deserve.
>
> **(Personal communication, April 2021)**

When Josh cofounded Flourish Veterinary Consulting in late 2017, it was built on the belief that the absence of stress and challenge is not necessarily the presence of thriving. The intention of applying positive psychology tenets in a tailored way to practice culture was to change the focus from simply eliminating or alleviating the known, work-related stressors to empowering leaders to cultivate an environment where informed associates were provided the structure and support needed to thrive. There were more subjective conditions that needed to be added to help individuals be their best selves—to move from the mindset of "just getting by" to the "thriving" state of being.

When I asked Josh what he found to be the biggest obstacle(s) to applying positive psychology to veterinary environments, he stated "the human brain". We discussed the way our human brains continue to operate in the outdated survivor mode of our ancestral hominids. Our brains and bodies respond to threats and obstacles in our lives, particularly at work, as challenges to positive change. Change is possible, but the complexities of the human environment and our biases need to be navigated when considering goals and strategies. Josh made great points: the need for patience, practice, fortitude, and optimism in order for us as individuals and as teams to lean into change and thus forward movement in our efforts to create the positive, collaborative culture that we desire in our practice environments.

Creation of metrics

- How to measure culture? How to track the human experience? After making desired changes in the practice, then how to measure the impact of these changes in meaningful ways?
- Need assessment tools! There are some tools that exist and can be adapted from human care environments, but likely new tools are needed to evaluate our veterinary environments more accurately.
- How do we define "success"? Profitability, retention, engagement versus associates thriving in their work? These are not mutually exclusive! Rather, if the humans are flourishing, the other metrics will naturally follow as will increased efficiency, productivity, and quality of work.

Dr Richmond was kind enough to contribute his own words of wisdom on/around the elements and impacts of positive psychology:

Depending on the source, skills of resiliency usually include the following: realistic optimism, gratitude, emotional regulation, causal analysis (flexible and accurate thinking), empathy, impulse control, and social support/reaching out for help (Reivich & Shatte, 2002)[7]. It is important to note, we all inherently have these skills—some are just better developed or come more naturally to us. While each of these skills are incredibly valuable, causal analysis and emotional regulation took the most effort to improve. Those two have also had the greatest impact on daily veterinary practice.

Regarding flexible and accurate thinking, it's been said that if our mind was an employee, we'd fire it. It makes connections that aren't there. Many times, we make significant decisions based on the faulty conclusions it makes. For example, let's think of a scenario. You text your boss. You really need to talk to her. She doesn't text back. Ever had the thought of "Dang, she must be mad at me?" Followed by, "it's probably because of that owner 3 weeks ago that talked to her". Now comes the negative spiral—loss of job, followed by loss of home, ending up living in my car with 3 dogs. Now, the ONLY fact in this scenario is that she didn't text back. You find out later she was helping a turtle cross the road and you just experienced a ton of negative emotion for nothing ... See what I mean? This skill can be summed up in one phrase: don't believe everything you think.

Effective emotional regulation has allowed me to respond to stressful situations fundamentally differently than in the past. This is where the breathing techniques Sonja has talked about are so effective (see Chapter 4 for the box breathing practice). First, being able to recognize the stress response in my body. Next, beginning to counter the sympathetic response by kicking the ol' parasymathetic NS with breathing exercises. It is almost habit now. What the recognition of the stress response sometimes lets me do is pause before responding to a client or coworker .... before saying or doing something I wish I hadn't. Being able to sidestep these mental land mines has had a truly positive effect on my wellbeing. I spend a lot less time ruminating over something I said or did.

Lastly, I would like to share about gratitude. For me, gratitude may be the most powerful tool there is to change my perception of the world around me. Josh spoke about negativity bias. Gratitude exercises counteract our negativity bias and prime our brains to spot good acts and intentions. The more "good" we see, the more likely we are to perform good deeds. This is a positive, upward spiral. In positive psychology, it is explained by Fredrickson's "broaden and build theory" (Author's note: BL Fredrickson, 2004, www.ncbi.nlm.nih.gov. The "broaden-and-build theory" describes the form and function of a subset of positive emotions, including joy, interest, contentment, and love). The positive emotion created by gratitude expands our scope of vision to look for good in the world. As a person in recovery, I can say that gratitude changes my perception of my condition faster than any substance. I don't say this lightly: gratitude is one of the reasons I am still here.

Skills of resiliency can be **learned**. This is the great news to share. Currently, we are implementing resilience training to the students at UFCVM and at conferences for the FVMA.

EUDAIMONIA

This is my new favorite word. I am sharing it with everyone I care about as such an important way to consider reframing your life, your work, your role on this earth in this lifetime. This term comes from the classic Greek philosophers (attributed to Plato himself) and literally translates as "good attendant" or "indwelling spirit". More recent and accurate translations have been proposed to be "human flourishing", "prosperity", and "blessedness". This is a much deeper and richer concept than mere happiness. It points again to the concept of "flourishing", of thriving, of living a life that is worthwhile, fulfilling, and elevating. This state of being is more deep-rooted and is connected to one's values and principles. In philosophical treatises by Aristotle on concepts like eudaimonia, he shared: "By living our life to the full according to our essential nature as rational beings, we are bound to flourish, that is, to develop and express our full human potential, regardless of the ebb and flow of our good or bad fortune". (*Nicomachean Ethics*)

In Japanese, the concept of eudaimonia is captured in the concept of *ikigai*, which means "reason for being". This sixth-century philosophy refers to having a meaningful direction of purpose in life, constituting the sense of one's life being made worthwhile, with actions taken toward achieving one's ikigai resulting in satisfaction and a sense of meaning in life" (Wikipedia, 2020). You likely recognize the French term "raison d'être" which also translates directly to "reason for being".

(Note: TED Talk November 27, 2018, Chloe Wong, "What is your Ikigai?")

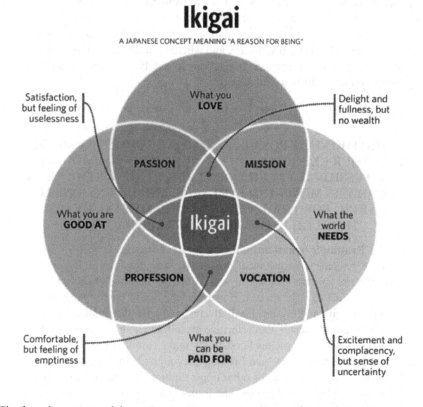

(The four dimensions of Ikigai. Source: Toronto Star, diagram by Mark Winn)

Over the last several years, much more has been written specifically about our veterinary community and profession exploring how and why focusing on the meaningful and fulfilling nature of our work is integral in shifting our individual and collective view of our work choices and lives. This conversation needs to start in the veterinary classrooms and continue into the practice environments, supporting meaningful and purposeful work, relationships, and opportunities for personal growth. Per Dr Jean Wallace,

> Meaningful work is important in understanding the wellbeing of veterinarians. Job characteristics (self-actualizing work, helping animals and people, and a sense of belonging) contribute to a sense of meaningful work, which in turn is related to eudae-monic wellbeing. Excessive job demands (work overload, financial demands and physical health risks) appear less relevant in understanding meaningful work but are clearly important in having negative consequences for veterinarians' wellbeing.[8]

The focus needs to shift to the many ways that practicing veterinary medicine and being a part of this profession provides deeply fulfilling opportunities to do something that you love, to be intellectually challenged, and to develop meaningful relationships. Creating a culture that supports individual flourishing, collaborative care for patients and for one another, and collective wellbeing and resilience is necessary and possible. A sustainable veterinary career is one where we know and remain tethered to our "why", where we as individuals develop buffering tools and knowledge to promote mental health and self-care, and where systems are fostered that promote sustainable work–life integration, psychological safety, learning, innovation, and progressive, ethical veterinary medicine.

EUDAIMONIA IS POSSIBLE! HERE ARE SOME EXAMPLES FROM OUR CLINICIAN COLLEAGUES WHO ARE MODELING PROFESSIONAL FULFILLMENT, CHERISHING TIME WITH FAMILY AS MOMS AND PARTNERS, AND INTENTIONALLY CARING FOR THEIR PHYSICAL AND MENTAL HEALTH THROUGH REGULAR SELF-CARE PRACTICES:

Dr Kate Brammer (small animal general practitioner, US):
For me, balance is a constant effort to make sure that I am giving my best to my family, my patients, and my myself. When something gets out of balance, I make an effort to right myself. However, when push comes to shove, I prioritize my family and myself. I will always be a wife and mom, and it is vital that I give them my best. I also prioritize myself. I'm no good to anyone if I am ill, tired, stressed. That means I constantly make sure that I am happy and am taking care of my needs. When I realize they aren't being met, I take steps to ask for what I need or look for other opportunities that may be a better fit for my long-term happiness. I am the key to my own happiness and my family's,

so I give my all, and if that isn't reciprocated by my situation, I take the steps needed to change it.

Dr Dani McVety (founder and business owner of Lap of Love Hospice, US):

Balance really came into my life when I decided to unapologetically live by my priorities. My family comes first, then my work, then everything else. As long as I feel like my home life is stable (getting my kids' schedules put together, making sure I have quality time with them and my husband, and so on), then my work mentality is clear, and I can focus on what I need to get done. I don't need to apologize to my kids for needing to work, or apologize to work for needing to spend time with my family. It's a decision I make clearly and on-purpose. And because I've made the decision that my family is top priority, I don't feel bad that I have to miss something at work or be delayed on an e-mail response. It just is what it is and it doesn't stress me out like it used to. I figure that at the end of my life, I will be much more happy with my investment in my family than anything else, so that's what I let guide me.

Dr Karen Fine (small animal holistic practitioner, acupuncturist, and graduate of Mindfulness Based Stress Reduction (MBSR) program; adjunct professor at the Cummings School of Veterinary Medicine at Tufts University):

What has been helpful to me is to *not* have a five-year plan and to be open to new interests and experiences. There have been times where I've needed to "coast" at work and times when I've embraced challenges in my career; both are okay and necessary. I'd also add to nurture your creative side, as it can help offset work-related stress.

(Note: Dr Fine just published the first veterinary textbook on narrative medicine, titled, *Narrative Medicine in Veterinary Practice*. Her interest and experience in this realm stems from many years of doing in-home visits, her experience as a holistic, integrative practitioner;

and her love of reading. She also has a forthcoming memoir, *The Other Family Doctor*, which is being published by Anchor Books.)

Dr Emma Whiston (BVSc [Hons]CHPV, certified hospice palliative care veterinarian, founder, My Best Friend Veterinary Home Euthanasia Service, Melbourne, Australia):

Being a veterinarian is an amazing vocation, but it certainly is not for the fainthearted. Having a veterinary degree can open the door to many different roles. Finding your niche can take time, experience, and creativity. Along the way, remember to seek joy, satisfaction, freedom, flexibility, autonomy, and integrity. Most importantly you must be responsible and proactive about your self-care; both mental and physical. Be kind to yourself. Create a team for yourself—this may include family and friends as well as colleagues, mentors, therapists. Sometimes you have to go through some crap in this job in order to find the awe. Patients may be difficult and their owners even more so, but you need to learn how to react positively to what happens to you. Strength and wisdom will come with experience but also by reaching out for help from your team when you need it.

Personally, now at 50 years of age, I am finally thriving and not just surviving as a veterinarian. Every day is different, never boring, and I feel (mostly!) equipped to deal with whatever the day throws at me. My team has helped me through many difficult times in particular with mental health issues, compassion fatigue, and burnout. I have built resilience and am no longer working from a place of fear.

I don't worry much about what competitors are doing, but I am very keen on healthy collaboration with colleagues as long as there is associated honesty, integrity and kindness. I am now running a successful practice with my husband, and as employers and mentors ourselves now, we practice my wise Grandfather's philosophy: "If you look after your staff, your staff will look after your clients, your clients will look after your business, and your business will look after you". We are part of our employees' teams, and we do our utmost to support them and let them know that they are not alone and that they can reach out to us for help at any time.

Dr Emily Weaver (associate equine practitioner [but about to start her own business], Florida private practice, US):
How does my life feel "just right" for me? I have been pondering this question, and I think that it can be very hard to balance the life of an equine vet and being a mom and wife, but for me it is all about practicing self-care and cultivating relationships. As an independent practitioner, I love having the freedom to set my own schedule and to practice medicine how I see fit. I still love horses, and I love seeing the relationships my clients have with their horses, but I "train" my clients! I teach them what is an emergency, what they can do on the farm until a vet can get there and teach them the best ways to care for their animals. I also am very clear that I can't be available to them 24/7 and to always call the emergency line when it is after hours. I have cultivated relationships with other equine vets in the area, so we can share the on-call responsibilities and cover for each other when we go on vacation. I schedule time in my day, almost daily, for some type of physical activity whether it be taking a walk, doing some yoga, or taking a kickboxing class! That is my "me" time and I know I am a better person for it. Most importantly is, I cultivate the relationships I have with family and friends that are my support system. Having a support system in place for those days when I can't make it to the bus stop in the afternoon or when I have a weekend filled with emergencies, I know I can call on them [to] help me with childcare, pet care, or whatever I need them to do; and I try my best to be there for them as well. I think I have the best husband and mom in the world, not sure I could do this without them!

Alyssa Mages (BS, CVT in small-animal care with over 17 years of veterinary caregiving, found herself drawn to peer-mentoring and creative,

novel ways to support training. Founder of Empowering Veterinary Teams, LLC): This journey, of life and vet med, is not one that has been straightforward, and I am all the better for it. I started on this path in the oceanic realm, and as I transitioned to a more terrestrial environment, I truly found my place. Starting on the bottom rung, working up and through a multitude of positions—technician, nurse, professor, coordinator, manager, business owner—I found not only my passion for medicine and educating my teams, but the continual evolution of myself. We cannot stop learning because life never stops teaching. So, I open my eyes, my mind, my heart. I listen, I grow, I become. As must we all for our profession, our industry, our world to continue to not only survive but to thrive.

NOTES

1. Cake, MA, et. al. (2015). The Life of Meaning: A Model of Positive Contributions to Wellbeing from Veterinary Work. *Journal of Veterinary Medical Education,* 42 (3), 184–93.
2. Wallace, JE. (2019). Meaningful Work and Wellbeing: A Study of the Positive Side of Veterinary Work. *Vet Rec.,* 185 (18), 571.
3. Roop, RG. (2004). "Compassion Satisfaction and Fatigue Survey". *The Humane Society of the United States (HSUS) 2003–3304.* Washington, DC: HSUS.
4. Rohlf, VI, et. al. (2021). Predictors of Professional Quality of Life in Veterinary Professionals. *Journal of Veterinary Medical Education.* (2021 June 8: e20200144. Doi.10.3138/jvme-2020-0144).

5. Volk, JO, et. al. (2020). Merck Animal Health Veterinary Wellbeing Study II. Survey results summarized and shared via online publication. https://www.merck-animal-health-usa.com/about-us/veterinary-wellbeing-study

6. Hamilton, N. (2019). *Coping with Stress and Burnout as a Veterinarian: An Evidence-Based Solution to Increase Wellbeing*. Queensland: Australian Academic Press.

7. Reivich, K, and Shatte, A. (2002). *The Resilience Factor: 7 Essential Skills for Overcoming Life's Inevitable Obstacles*. New York: Broadway Books.

8. Wallace, JE. (2019). Meaningful Work and Wellbeing: A Study of the Positive Side of Veterinary Work. *Vet Rec.*, 185 (18), 571.

6 Get Excited and Inspired! The Veterinary Profession of the Future

Be curious, not judgmental ...

—Walt Whitman

We cannot solve our problems with the same thinking that we used when we created them.

—Albert Einstein

By celebrating our colleagues' creative endeavors to shift the dialogue, we can be inspired and motivated to engage in systemic change. As a result, we can collectively and intentionally reshape our professional identities and ecosystems to support thriving veterinary professionals in sustainably healthy environments.

I am hopeful that you now understand and see the value of positive reframing of the entire discussion supporting our profession's future. This chapter is dedicated to further helping you to shift into a brain space with both optimism and hope. We find ourselves today as a profession, and as a global human community, faced with complex challenges in many realms that impact our lives. There is a power surge of creative energy as a result of positive change agents around the world. There is active work seeking new approaches and solutions to long-standing dysfunctions and inequities. In veterinary medicine, there is a more focused attention by our professional community on how best to develop and foster collaborative, compassionate initiatives that sustainably support the veterinary professionals of today and of tomorrow. These efforts will benefit all of us and result in better care for our veterinary patients and the client community that we serve.

Here, I am excited to shine a spotlight on a sampling of the diverse ways that our community is showing up around problem-solving, wellbeing support, and thoughtful peer-based mentoring. There is no way that I could possibly capture all of the amazing initiatives that exist or are being developed, here in this chapter, but I am seeking to celebrate the hard work and creative energy being invested by our fellow veterinary colleagues. I want you to know what is going on in our profession and have a sense of pride and perhaps be inspired yourself to contribute. Where can you participate? What speaks to you? What might be overlooked that perhaps you might be uniquely qualified to move that specific conversation forward?

Do not discount the power of your voice, your perspective, or your intuition. We all must be courageous and committed. Developing a workplace that supports both

DOI: 10.1201/9780367816766-6

the wellbeing of the patients and the caregivers requires a purposeful shift in perspective toward realistic optimism and informed hope.

THE WRITTEN WORD

In the last 15 years, there has been a positive, progressive trend of increasing research efforts specifically investigating a myriad of issues that impact veterinary professionals. The resulting articles and books that have been written come from around the world and address topics ranging from mental health concerns, moral stressors, ethical dilemmas, substance use disorder, and personality traits common to veterinary caregivers (e.g., imposter syndrome), to evaluation of training structure and curriculum content. In the last eight years, there has been an uptick of both research-based and experience-based articles written as well as blogs, online continuing education (CE) and resources, and books all addressing varying aspects of veterinary wellbeing. The qualitative data and metrics from research will support strategic decision-making as well as the creation and implementation of meaningful solutions. Additionally, the research efforts and results combined with the broad-spectrum of interest in holistic veterinary wellbeing are normalizing the conversation overall. These are significant shifts and will contribute positively to the efforts from many fronts to address long-standing issues in our profession. I have included a small smattering of articles and books in the Recommended Readings portion of this book to pique your interest.

CREATIVE ENDEAVORS AND THE SPOKEN WORD

Creature Conserve—Conservation and Education through Art

Dr Lucy Spelman (DAZVM) is working with artists to bring life and connection through art/education to wildlife and conservation efforts. The mission of **Creature Conserve** is to "bring artists, creative writers, and scientists together to foster informed and sustained support for animal conservation". Their efforts include art exhibits and workshops, field studies (global), network growth, and scholarships. Check out www.creatureconserve.com and Dr Spelman's website: www.drlucyspelman.com.

Web-Based Community Support and Virtual Learning for Veterinary Communities

PODCASTS

SO many goodies to listen to, but here are several to get you started:

- **The Vet Reset**. Hosted by Dr Katie Berlin an associate veterinarian in central Pennsylvania who, through (her) own battle with burnout, developed a passion for the role physical wellness plays in mental health and happiness. Her goal with The Vet Reset is to be a resource for other vet professionals to find encouragement, community, personalized and practical advice, and a hefty dose of inspiration to find their way to their healthiest and happiest selves. (https://thevetreset.net)

- **The Joyful DVM Podcast** Hosted by Dr Cari Wise. Started in May 2020, Dr Wise's coaching platform and podcast share her desire to "empower veterinary professionals" and address issues, such as the high levels of mental health concerns which some may view to be inevitable consequences of VetMed careers. "Different outcomes require a different approach and perspective". (https://joyfuldvm.com)
- **Podcast a Vet**. Hosted by Dr John Arnold, Podcast A Vet is a podcast for veterinarians, students, nurses, veterinary professionals, and animal lovers with an emphasis on community. The podcast shares the stories, struggles, successes, and insights of leaders across the veterinary industry. (https://podcastavet.com)
- **RadioVetNurse.** Hosted by Cat Robinson, a Cert IV VN out of Australia who has had the opportunity to experience many different facets as a veterinary nurse and practice manager. Radio Vet Nurse is an interview show where she hosts a different vet nurse in each episode to create space and dialogue around the challenges, strategies, and successful solutions these veterinary professionals have experienced in their career. (https://www.radiovetnurse.com)
- **The Whole Veterinarian.** Hosted by Dr Stacey Cordivano, The Whole Veterinarian podcast is about sharing experiences and starting a dialogue about how we can grow as people and professionals. As a female equine solo practitioner, mother, and business owner, Dr. Cordivano states on her website that she has seen and can relate to the unique challenges that someone in her boots is experiencing and the challenges that face them. With that in mind, her podcast seeks to acknowledge and support that "Vets are some of the hardest working and most dedicated professionals out there; together, let's learn how to develop a life we love." (https://thewholeveterinarian.com)
- **Have you Herd?** Brought to you by the American Association of Bovine Practitioners, an international association of cattle veterinarians and veterinary students dedicated to the health, productivity, and welfare of cattle. (https://aabp.org)
- **VetCandyLIFE.** Hosted by Dr Quincey Hawley and Renee Machel (LVT). Each episode features expert tips, lifestyle advice, and real-life experiences from the most exciting veterinary professionals in our industry. (https://vetcandyloife.podbean.com)
- **TheVetpodcast.** Hosted by Dr Bryan Gregor from New Zealand. (www.vetpodcast.weebly.com)
- **BluntDissection.** Hosted by Dr Dave Nicol. The best minds in veterinary medicine, academia and business profiled so you can learn from their experience. (www.drdavenicol.com)

FORUMS FOR DISCUSSION, CE, AND "TOOLBOX BUILDING"
FOR ALL VETERINARY PROFESSIONALS

VETGirl. Founded by Drs Garrett Pattinger (VMD, DACVECC) and Justine Lee (DVM, ACVECC, DABT), VETGirl is a subscription-based multimedia service offering RACE-approved online CE of all varieties for veterinary professionals. Their social media is a resource of humor and knowledge as well as supportive space for the veterinary community. Their team is also now offering VETGirl Veterinary Continuing Education Podcasts.

KickAss Vets. Founded by Dr Ann Herbst, a delightful and truly informative website that provides information and tools for veterinary associates to take control of their careers as well as professional and personal wellbeing. Herbst and her team provide webinars, consulting services, and regular sassy and strong blogs to "remind and motivate people to find themselves again, find that passion and drive again, and unleash their inner KICK ASS VET!" (https://kickassvets.com)

The Bridge Club. Founded by Catherine Haskins. Having worked within veterinary medicine for over 20 years, Catherine created a video-based community allowing all professionals to network remotely, create real and authentic connections to help advance the profession, develop new partnerships, and grow their businesses. (www.thebridgeclub.com)

Veterinary Visionaries. A future-focused collaboration where veterinary professionals, organizations, and industry partners can create novel approaches to existing and future challenges. (www.aaha.org)

CURRICULUM AND CLASSROOM CULTURE CHANGES: DREAM TO MATCH THE NEED

IN VETERINARY TRAINING

There are a multitude of opportunities to shift veterinary education toward a more positive, supportive, individualized, and real-world framework. The myriad of concerns that currently exist in the veterinary learning environment are diverse and complex, requiring both creative and responsible discussions.

- There are several veterinary colleges that were early adopters of incorporating soft-skill development into their four-year curricula (communication skills, leadership development, wellbeing topics and skills, ethics).
- Evaluation of clinical content as well as delivery/teaching modalities are underway to adapt to the marked increase in medical knowledge that one needs to be considered a prepared, competent veterinary practitioner (within their chosen veterinary field).
- There are colleges that have adopted the hybrid distributive model of creating collaboration between academic teaching hospitals and clinical partners in their communities.

- Embracing of new technologies and virtual teaching platforms is in its early days but is being increasingly explored and utilized.
- The awareness of the need to shift our professional culture is being supported with early discussions around increased inclusion and diversity, increasing psychological safety to enhance wellbeing and learning, and decreasing stigma around mental health challenges experienced by many of our colleagues during all stages of their professional caregiving journey.

Here are just a few examples, but please know that there are many more out there. Get curious and contribute to the conversation as well!

1) **Association of American Veterinary Medical Colleges (AAVMC)**
   Development of competency-based framework (learn more at http://bit.ly /2BnxktX). Per an article by Dr Susan Matthew in October 2020 (JAVME, vol.47, issue 5, pp. 578–93), "competency-based medical education is an educational innovation implemented in health professions worldwide as a means to ensure graduates meet patient and societal needs". The CBVE framework consists of nine domains of competence and 32 competencies; institutions adopting this framework would have greater capacity for outcome assessment analysis. The nine domains of competence include clinical reasoning and decision-making, individual animal care and management, animal population care and management, public health, communication, collaboration, professionalism and professional identity, financial and practice management, and scholarship. The ability to prioritize skills, digitize learning and competency tracking, and create options around time spent in the traditional education environment could allow for increased flexibility for the how, when, and expense associated with the learning process.

2) **University of Florida College of Veterinary Medicine (UFCVM)**
   - UFCVM had initiated a four-year extracurricular wellbeing education program with assistance from Banfield. This program was initiated by Drs Juan Samper and Amanda House of UFCVM, with consultation from Scott Mogren and Drs Paul Dutcher and Seth Vrendenberg from Banfield and Dr Philip Richmond from the FVMA Professional Wellness & Wellbeing Committee.
   - The curriculum consists of the induction of positive emotions, purpose work, strengths assessment, resilience skills training, communication, emotional intelligence, and psychological safety/positive workplace culture. There are plans for psychological safety and positive culture in the workplace training in the area of clinical instruction.

3) **Cummings School of Veterinary Medicine at Tufts University**
   - *Tufts Healer's Art Elective Course* was started in Spring 2017. The course now consists of five 3-hour sessions and emphasizes experiential learning. Topics have included mindfulness training, embracing stress, and time management.

The Healer's Art Course (https://rishiprograms.org/healers-art/) is an innovative discovery model course in values clarification and professionalism taught in both medical and veterinary schools around the world. The course was tailored for veterinary students after Drs Jane Shaw, Camille Torres-Henderson, and Laurie Fonkin attended the workshop and they introduced the course to Colorado State University veterinary students in 2002. Now, over 35 veterinary faculty from 15 schools have taken the faculty development training and eight schools have added the course to their curriculum.

- *Ethics in Clinical Practice Elective Course* was created and supported by Dr Lynn Roy. The focus is to discuss ethical dilemmas and increase ethical literacy to small groups of fourth-year students. Her description of the course's intent to the students:

  Over the past three years, you have had an ethics curriculum that included the human–animal bond, veterinary medicine and the law, euthanasia, and veterinary medical ethics. As with all of your training, fourth year is when we get to put it all into practical application. Here, we get to discuss ethical cases, who the stake holders are, and what options are to be considered … This is all to be done to give you some "tools" for your tool belt as you go forward and within the next year and soon become the veterinarian responsible for these cases.

- Development of novel elective courses that support "soft skill" development and critical thinking, e.g., *Narrative Medicine* with Dr Karen Fine. Her upcoming book, *Narrative Medicine in Veterinary Practice*, is the first veterinary textbook to explore the application of medical humanities through narrative medicine in veterinary medicine. (Anticipated publication date September 2021.)

4) Royal Veterinary College of Veterinary Surgeons
- The Directorate of Learning and Wellbeing was created in 2017 with the aim of promoting wellbeing, learning, and working environments for all RVC students and staff that are more inclusive and accessible. Per the RVC website, it works to create opportunities, events, spaces and learning experiences which contribute to a safe and respectful community across the RVC.
- RVC Blended Learning uses new ways of learning and interacting through technology to support the best of their live-taught course with innovative online learning. Per their website, this allows for the student to learn at their own pace and to fit the course around work and family commitments. (www.rvc.ac.uk)

5) University of Bristol—Veterinary School
- Per Dr Lucy Squires, a teaching associate and one of the wellbeing educators at Bristol Veterinary School, there is a wide variety of courses in

all four years that helps students to promote mental health knowledge and skills to support self-management and thriving in and after their veterinary school experience. Peer–mentors lecture on wellbeing topics as well as preparation for clinical practice topics (e.g., communication, time management, conflict management).

- The *Mental Wellbeing Toolbox Handbook* is freely available online at www.bris.ac.uk/vetscience/media/docs/mental_wellbeing.pdf. The handbook includes details of all topics covered in the Mental Wellbeing Toolbox, including a complete list of references from the literature review. There is a small section on specific resources available at the University of Bristol, but there are similar services available in the other vet schools and more widely. The handbook can be used as a template by others considering putting together a similar document (although we request the University of Bristol be fairly credited in this case). We'd also welcome questions via e-mail (lucy.squire@bristol.ac .uk) from institutions, organizations, or practices interested in taking a similar approach.

6) **"Radiology Rules"**—Virtual CE platform developed by Dr Betsy Charles:

I started a company that is dedicated to rethinking education and what it could look like not only in radiology, my specialty, but also in the leadership and DEI spaces. I like to take these sometimes vague, often lofty and really challenging topics and break them down into practical, bite-sized, easy to digest morsels of awesome that are useful. I DO this because I want people to actually put what they have learned into practice. Like right away! I want them equipped to make a difference immediately. That's how you change the world—one bite-sized morsel of awesome at a time.

PROFESSIONAL DEVELOPMENT FOR VETERINARY TEAMS

- **Empowering Veterinary Teams (EVT)**—cofounded by Alyssa Mages (a 17-year veterinary technician veteran and creative, teaching powerhouse) alongside Caitlin Keat (engineering and organization guru for EVT) as her COO in 2020. EVT's goal is to inspirate and impact the veterinary profession's evolving culture change through the creation of novel live and virtual trainings to support professional development of all veterinary associates, particularly the nursing staff. Their mission is to "be advocates of growth and empowerment to provide the highest standards of veterinary care to create positive and dynamic work environments".

   (https://empoweringveterinaryteams.com)

- **GetMotiVETed Wellbeing Solutions**—cofounded by Dr Quincy Hawley and Renee Machel.

   This wonderful, dynamic duo are passionate about supporting veterinary professionals through their blogs, podcasts, and speaking engagements. They have created GetMotiVETed University as a CE hub and recently coauthored ClinLife-21 with the EVT team (above).

The vision: To be a global leader in ensuring veterinary wellbeing for all members of the veterinary community. (www.GetMotiVETed.com)

- **Veterinary Team Training (VTT)**—Amy Newfield, MS, CVT, VTS (ECC). Amy is a very experienced and humorous mentor and internationally known speaker and writer for veterinary teams with a passion for creating healthier and happier teams, empowering professional development, and "finding unicorns". Her new book, *Oops! I Became a Manager: Managing the Veterinary Hospital Team by Finding Unicorns* is now available (2020). (www.vetteamtraining.com)

Virtual Conferences and Learning

- **Uncharted Veterinary Conference**—Founded by Dr Andy Rourk. From the Uncharted website, the goal of this initiative is to be "veterinary medicine's premier business and career development community. The goal of Uncharted is to help you thrive in everything from being fantastic in the exam room to dealing with practice drama to getting more done during business hours so you can go home and recharge". (https://unchartedvet .com)
- **VetCandy**—Founded by Dr Jill Lopez. This organization and the associated website supports many resources (written and video) as well as links to their podcast, "TV", and CE endeavors/allies. (https://myvetcandy.com)
- **VetBloom**—As described on their website, "VetBloom's creation was rooted in the training needs of a specialty and emergency veterinary group in New England. "Using industry experts, VetBloom has developed a variety of curricula, continuing education, and skill protocols that have, in turn, effectively contributed to the growth of a wide audience of veterinary professionals". Their virtual offerings include content for all veterinary teammates and their wide range of RACE-approved CE subjects can be engaged with as individual courses or as a complete curriculum. (https://vetbloom/ com)

THE NOVICE PROFESSIONAL EXPERIENCE AND PEER-MENTORING

Peer-mentoring is an exciting opportunity for us to positively impact individual new veterinary graduates and the overall future of the veterinary profession. New initiatives are being developed to support the "novice practitioner" in both clinical and life skills to fortify their ability to thrive in both their personal and professional lives.

A wonderful example is a program being developed currently by Dr Addie Reinhard with the support of Merck Animal Heath. Dr Reinhard's program is intended to compliment the classroom curriculum with a focus on the support of the novice practitioner period when the real-world application of academic learning meets the numerous challenges of the clinical environment. This is her description of the program that she created:

The mentorship and professional development program I have created is for new and recent veterinary graduates. This evidence-based transdisciplinary program incorporates structured peer support and training in professional skills to support new veterinary graduates during the transition to practice. Early results show that veterinarians who participated in the program had lower levels of exhaustion, a component of burnout, after the program as compared to veterinarians who did not participate in the program. By providing veterinarians with resources and training in the transition to practice, we hope to reduce stress and burnout and improve wellbeing during the start of the veterinary career. (https://mentorvet.net)

EI & D EFFORTS—INCREASING AND SUPPORTING DIVERSITY IN THE VETERINARY PROFESSION

- **Pawsiblitities Vet Med**—Founded by Dr Valerie Marcano (DVM, PhD) and Dr Seth Andrews (PhD). A non-profit organization that aims to improve the recruitment and retention of underrepresented groups in the veterinary profession (www.pawsiblitiesvetmed.com).
- **BlackDVM Network**—Founded in 2018 by Dr Tierra Price while still a veterinary student. The organization is intended to serve as a platform for the empowerment of Black veterinary professionals. Their mission is to create a community for Black veterinary professionals to grow, connect, and advance in veterinary medicine. For additional information and resources, check out their website at www.blackdvmnetwork.com.
- **PrideVMC.** The mission of PrideVMC is to create a better world for the LGBTQ+ veterinary community. (www.pridevmc.org)
  - **LGVMA**—Founded in the US in 1993 (born of several other organizations starting in 1977 to support the LGBT community). *Rebranded in 2018 to become Pride VMC.*
  - **Australian Rainbow Vets and Allies (ARVA)**—Founded in Australia in the 1990s.
  - **British LGBT+ Veterinary Group.**
  - **Queer Vets of Germany.**
- **BSVSA (Broad Spectrum Veterinary Student Association)**—Founded in 2010 by veterinary students at the UC Davis SAVMA Symposium. The BSVSA mission is to connect, support, and empower community for LGBTQIA students and allies across veterinary education. For further information or if you have any questions, e-mail: broadspectrumoutreach @gmail.com.
- **Veterinarians as One Inclusive Community for Empowerment (VOICE)**—This student-run organization supports diversity and inclusion in a wide variety of ways on campus, by providing leadership and mentorship to youth interested in veterinary medicine careers, to meeting the needs of diverse clientele. At the time of this book-writing, VOICE is the EI & D organization that has the largest number of chapters in the US and island veterinary schools.

ORGANIZATIONAL SUPPORT

Examples at the national level: AVMA, CVMA, BVA, AVA.

Each organization's website provides a wide variety of resources (financial advice, CE opportunities and links, social medial/cyberbullying guidance, legal advice, wellbeing resources, etc.). These resources are invaluable for veterinary professionals at all stages of their career (pre-vet, in training, recent graduates, and practicing professional support/development for veterinary technicians and clinicians).

Examples at the Global level:

(WSAVA) World Small Animal Veterinary Association. (https://wsava.org)
(WVA) World Veterinary Association. (www.worldvet.org)
Veterinarians International. (https://vetsinternational.org)
One Health Initiative: **One Health** is a collaborative, multi-sectoral, and transdisciplinary approach—working at the local, regional, national, and global levels—with the goal of achieving optimal **health** outcomes and recognizing the interconnection between people, animals, plants, and their shared environment. (www.cdc.gov/onehealth)

ORGANIZATIONAL CHANGES IN HOUSE OFFICER TRAINING STRUCTURE/CONTENT/EXPERIENCE

Currently, there is an increased focus on wellbeing support and development while continuing to provide strong clinical knowledge and skill development opportunities in many training programs. There are several universities and private practice organizations working alongside veterinary social workers and in-house mental health professionals to develop what may be templates for others around updated and healthier training programs and culture. By incorporating wellbeing CE and facilitated discussions throughout the training period, the goal is to improve the experience for house officers and mentors alike. Systemic change around scheduling, structure of training time, and compensation is also being concurrently evaluated. Novel approaches to traditional internships, such as accelerated emergency medicine training programs (e.g., EmERge from BluePearl and the Emergency Mentorship Program with VEG) provide important opportunities for the desired peer-mentoring those new graduates are seeking.

HEALTH AND WELLBEING CERTIFICATIONS

AVMA Workplace Wellbeing Certificate Program is an excellent way to contribute to one's individual understanding of veterinary wellbeing. The lectures provide training ranging from examination of mental health concerns, strengthening communication skills (e.g., giving/accepting feedback), and fortifying a safe work culture and team ethos (the equality/inclusivity/diversity lecture), to increasing confidence around supporting necessary conversations on psychological distress and suicide (via the QPR certification). Building upon this work, there is currently an initiative to create a platform that would support wellbeing in the workplace by training individuals to lead these conversations and provide the resources needed. Anticipated launch some time in 2021. (www.axon.avma.org)

Mental Health First Aid (MHFA).

An international program to teach the skills to respond to the signs of mental illness and substance abuse. Started in Australia in 2001, the program came to the US in 2008 and to several other countries around the world since that time. This program has been taught and is actively being used in the veterinary communities in Australia and the UK. Now, the MHFA is being used more broadly in the US for healthcare workers, including those in the veterinary profession. (www.mentalhealth firstaid.org)

Green Cross Traumatology Academy.

The Green Cross Academy of Traumatology (GCAT) website, the initiative "was initially organized to serve a need in Oklahoma City following the April 19, 1995 bombing of the Alfred P. Murrah Federal Building. In 2004, GCP merged with Green Cross Foundation and emerged as the Green Cross Academy of Traumatology. We are an international, nonprofit, humanitarian assistance organization comprised of trained traumatologists and compassion fatigue service providers. Most are licensed mental health professionals; all are oriented to helping people in crisis following traumatic events.

There are a variety of workshops and courses available, such as the Compassion Fatigue Educator which can lead into the certification of Compassion Fatigue Therapist. On their website, there exist many other additional certifications around supporting traumatized individuals in the field (disaster scenarios) and in direct practice. (www.greencross.org)

PROGRAMS WITH CURRICULUM CONTENT FOR ALL CAREGIVERS

- Master of Applied Positive Psychology (MAPP) and Coaching Psychology hybrid online/on-site programs offered at: University of Pennsylvania (www.lps.upenn.edu/degree-programs/mapp); University of Melbourne (https://study.unimelb.edu.au); University of East London (https://www.uel.ac.uk); and Anglia Ruskin University (https://www.arc.ac.uk).
- Certificate in Applied Positive Psychology (CAPP), The Flourishing Center (both online and virtual programs available). All teaching faculty are graduates of Penn's MAPP program. (https://theflourishingcenter.com)

- Coursera's courses (https://www.coursera.org/courses)
  - Foundations of Positive Psychology Specialization (taught by Penn's MAPP faculty).
  - Coursera's Science of Wellbeing (taught by Yale faculty). Highly rated course.
  - Coursera's Foundation of Mindfulness.
  - Coursera's Psychological First Aid (taught by Johns Hopkins faculty).
- EdX's Science of Happiness course. Taught by UC Berkeley's Greater Good Science Center faculty. (https://www.edx/org/course/the-science-of -happiness)
- EdX's Science of Happiness at Work professional certificate. Learn skills to boost satisfaction, engagement, and collaboration in workplace settings. (https://edx.org/professional-certificate)
- Certified Compassion Fatigue Professional (CCFP). (https://www.evergre encertifications.com/evg/detail/1007/certified-compassion-fatigue-profes-sional-ccfp)

U Penn's The Flourishing Center (theflourishingcenter.com) is one of several institutions offering this program. Per their website,

Applied Positive Psychology is a discipline that examines the intersections of body, brain, culture, and science to develop tools and practices that enhance human flourishing and wellbeing. In this certificate program, you will be introduced to the field of positive psychology and will learn tools and practice strategies that support personal, organizational, and community wellbeing. The online courses in applied positive psychology teach you the theoretical and empirical foundations of human flourishing, how wellbeing is measured, and what activities increase human flourishing in various contexts and settings.

US Army (Impact on Army Veterinary Corps): The connection between wellbeing and performance is supported by the October 2020 release of the Army's training and field manual that is intended to support "peak performance". In this latest revision (the first in 8 years), titled the "FM 7-22 Holistic Health and Fitness" manual, there are segments on sleep hygiene, meditation, serving others and "spiritual readiness". These are intended to support the individual in "the development of the personal qualities needed to sustain a person in times of stress, hardship, and tragedy. These qualities come from religious, philosophical, or human values and form the basis for character, disposition, decision making, and integrity.""

LEADERSHIP DEVELOPMENT WITH CULTURE IN MIND

Vet Integration Systems (VIS) was founded by Dr Ivan Zak, a long-term small animal emergency clinician. Dr Zak's MBA dissertation focused on "Lean Thinking in Veterinary Organizations to Improve Employee Experience". "Lean leadership" is being considered as a model to address burnout in veterinary professionals and increase collaborative communication, resulting

in engagement and compassion satisfaction (Organization Excellence—Veterinary Integration Systems—Lean Management concepts). VIS is a consulting team but also offers podcasts and virtual informative panel discussions including both veterinary professionals and outside subject matter experts on a wide variety of impactful topics supporting successful veterinary practices and associates. (https://vetintegrations.com)
Women's Veterinary Leadership Development Initiative.
As stated on their website,

"The veterinary profession has undergone a great demographic shift over the past 25 years, but the change in leadership hasn't kept pace. Women, now more than half of the profession and rising, need to be better represented in leadership roles if we're to deliver on the promise of veterinary medicine to society. WVLDI is dedicated to helping develop those leaders—in organized veterinary medicine and from practice owners to academia to corporate boards—and to achieve a stronger, more effective, and personally rewarding profession."

Their vision: *"to inspire and empower veterinary leaders to harness diversity and equity through creating and inclusive professional community where all members can deliver on the promise of the veterinary oath"*. (https://wvldi.org).
Flourish Veterinary Consulting, cofounder and lead consultant, Josh Vaisman (CCFP, MAPPCP).
Josh's optimism and many years in the veterinary profession informed his pursuit and his degree in Positive Psychology and Coaching Psychology. He and his team provide consulting services to practice, interactive lectures and workshops, keynote addresses, and leadership coaching. (www.flourishveterinaryconsulting.com)
VetResults was founded by Pam Stevenson, a certified veterinary practice manager with more than 38 years of experience in a variety of practice environments. Pam certainly has experience and wisdom to share! She provides in-depth consulting with practices and teams to allow for customized, practical solutions to reach practice goals around improved culture, productivity, and profitability. (https://vetresults.com)
CatalystVet Professional Coaches, founded by Rebecca Rose, CVT, and created to support veterinary team members in creating a healthy, sustainable life and career in veterinary medicine. The offerings through Catalyst are abundant, and many are free, such as forums to support mindfulness and gratitude, VetPC Blog and Vlog, virtual CE at conferences, and a monthly publication in the Veterinary Practice News on topics such as communication skills, team building, and career advancement. (https://catalystvetpc.com)
Veterinary Leadership Institute (VLI) and the VLE program.
Dr Betsy Charles (DVM, MA) is a courageous entrepreneur and one of the founding voices in the conversation around veterinary professional wellbeing. Her optimism and tenacity bolstered her during all her years of working with the VLI. The VLI is a "non-profit organization dedicated

to the development of healthy and resilient leaders who can make a positive difference for the profession" (from the VLI website: https://veterin aryleadershipinstitute.org). The programs that have blossomed out of the VLI, such as the Veterinary Leadership Experience (VLE) and the Student Leadership Summit, have fortified many professionals in different stages of their career to find their voice in contributing to the positive change they may wish to see in the veterinary profession. I recently asked Dr Charles to share a bit more with me about her personal view of the VLI/VLE today:

Most mornings these days I wake up and I am wildly optimistic about what the day will bring, and I can't wait to shine light and get after it. Every once in a while, though, I wake up and all I see facing our profession (and, interestingly, me personally) is a huge, complicated mountain that will require mad climbing skills in order to reach the summit. Like I said, most days I am totally game for the hike and feel prepared for the road ahead, but sometimes I want to sit at the base of the mountain, head in hands, and lament what it's going to take to get to the top of the mountain. It's this tension that I am now intimately familiar with, the tension between the good and the terrible. I want to see the view from top, but I don't necessarily want to do the work necessary to get there. We can pretend this tension doesn't exist. We can try to avoid it. We can do what veterinary professionals love to do: we can just keep climbing without paying attention to the skill sets that will allow us to get to the summit safely and in one piece.

Or, we can learn how to become expert climbers.

Enter the Veterinary Leadership Institute! This organization, or more importantly, the people within this organization, are expert climbers. We are built on a foundation of servant leadership and wellbeing and are devoted to creating programs that equip veterinary professionals to make the climb in an efficient and effective manner. What makes us different, though, is our commitment to the intensity of the training necessary to get to the top. It is not easy and requires real personal work that happens in the context of community, but the view from the top is so worth it.

The Veterinary Leadership Experience (VLE) is a big part of why I wake up most days wildly optimistic. When you are prepared and ready to face the tough terrain in front of you and you have a team of equally skilled climbers next to you, navigating the tension between the difficulty of the climb and the view from the top is much less daunting.

GLOBAL VETERINARY WELLBEING CE EFFORTS

CANADA

- **Dr Marie Holowaychuk** (DVM, DACVECC) is a veterinary wellbeing force phenom! She has pursued additional certifications in Compassion Fatigue Training, Mental Health First Aid, Applied Suicide Intervention Skills Training (ASIST), and her 200-hour yoga and meditation training. Over the years, Dr Holowaychuk's support and available resources around veterinary professional wellbeing have grown from speaking engagements, workshops, and retreats to online courses, webinars, and coaching opportunities. Explore her website at https://marieholowaychuk.com.

- **Dr Debbie Stoewen** (DVM, MSW, RSW, PhD) is the CEO (care and empathy officer) of Pets Plus Us and director of veterinary services. She is an academic, entrepreneur, and facilitator, committed to advancing the health and welfare of people and animals at the intersections of industry, academia, and civic society.With 25 years of practice experience, including being the founder and sole proprietor of a companion animal hospital, she is well-versed in the daily realities of veterinary practice. She adds to this an advanced understanding of personal wellbeing, interpersonal dynamics, and veterinary-client-patient communication, giving her a unique vantage point on the social aspects of practice. She has many published articles and podcast interviews available online and is an active lecturer at veterinary conferences in the US and Canada.

AUSTRALIA

- **Positive Psych Solutions**—Founded Dr Nadine Hamilton. Dr Hamilton is a clinical psychologist and has completed her doctoral research into veterinary wellbeing. She has had a vested interest in supporting veterinary professionals to "get on top of stress and psychological fatigue to avoid burnout and suicide". She is an internationally known speaker, author, and contributor to a wide variety of veterinary professional wellbeing initiatives. (https://positivepsychsolutions.com.au)
- **Make Headway**—founded by Dr Cathy Warburton. After 25 years as a clinical veterinarian and manager, Dr Warburton completed her degree in Positive Psychology and Wellbeing and further training coaching which led to her starting Make Headway. Cathy shared the following summary of her professional wellbeing and positive culture platform:

    Goals: At Make Headway, we are veterinary professionals empowering other veterinary professionals to understand themselves, so that they can build the good, negotiate the challenges, and craft a satisfying, meaningful, and sustainable career.

    Initiatives: We work compassionately and collaboratively, each bringing our own individual and complementary experiences and skills. We are passionate about one-on-one, small group, and work-place coaching and debriefing.

    Professional passions: We walk alongside our clients, providing personalised support. We love to work with what is in the room and what is important right now. (https://makeheadway.com.au)

- **Dr Vanessa Rohlf,** BA(Hons), MCouns&PsychTh, PhD. Vanessa is a registered (licensed) counselor and psychotherapist who specializes in providing compassionate, tailored, and evidence-based support to those who care for animals. She specializes in a range of areas, including compassion fatigue, burnout, stress management, mindfulness as well as pet loss and grief.

Vanessa is a former veterinary nurse and draws on her experience and qualifications to support veterinary professionals in three major areas. First, she works to promote personal and professional wellbeing and resilience. Second, she works with clinics to help them provide compassionate emotional support to clients experiencing trauma and grief. Third, she is a qualified and experienced referral source for clients requiring one-on-one tailored counseling for emotional support. (http://drvanessarohlf.com.au/)

United Kingdom

- **The Mind Matters Initiative (MMI)** is run by the Royal College of Veterinary Surgeons and aims to improve the mental health and wellbeing of veterinary team members. The offerings from MMI range from live and virtual training/CE, wellbeing resources, involvement in research endeavors, and more. (https://www.vetmindmatters.org)
- **Dr Katie Ford BVSc** (Hons, CertAVP (SAM), PGCert MRCVS). Veterinary surgeon, speaker, and "imposter buster", Dr Ford offers coaching, group coaching, online CE training, and mentorship. Check out her website to learn more about her and her offerings. (https://www.katiefordvet.com)

United States

- **Julie Squires, BA, CCFS** is a compassion fatigue specialist and certified life coach. She created her professional platform to support the health and thriving of veterinary professionals and teams in 2015. Her offerings have grown to include a wide variety of webinars, articles, team training and individual coaching opportunities, as well as a great podcast. Julie clearly has a heap of passion and knowledge to bring to these important conversations and may be the right resource for you or your practice as you grow and evolve. (https://www.rekindlesolutions.com).
- **Jessica Dolce,** educator and coach. Jessica has been developing her resources and trainings to transform compassion fatigue and fortify resilience for veterinary professionals and has spent over 20 years working with or around companion animals and veterinary professionals in a variety of ways. Witnessing the negative effects of Compassion Fatigue motivated her to obtain her Certified Compassion Fatigue Educator (CCFE) via The Green Cross. She has a variety of other certifications in/around coaching, positive psychology, and mindfulness as well. Her online platform "Practice Compassionate Badassery!" and associated resources are very worth exploring! @ https://jessicadolce.com

SOCIAL WORK SUPPORT FOR THE VETERINARY PROFESSION

There has been an exciting and much-needed increase in the presence of licensed social workers, and now veterinary social workers, working in veterinary teaching hospitals and practice environments. Historically, these individuals were hired to largely support the needs of clients, providing primarily animal-related grief and bereavement services. Over the past 10 years in particular, there has been a gradual increase in mental health professionals working directly with students, staff, and faculty in the veterinary schools and with all associates in the private practices. Regular wellbeing rounds, providing resources and knowledge to support mental health in challenging work environments, and creating a safe space for compassionate listening and support are becoming much more common. Referral to appropriate additional mental health professionals and groups are facilitated as needed for the individual seeking support.

I want to spotlight some of the "super star" social workers that have been, and are, instrumental in the work that has led to current momentum to have social worker support in many more veterinary spaces. Additionally, these dedicated women are significantly contributing to the shift in our profession's culture toward one of improved physical and mental health of the caregivers:

- Dr Elizabeth Strand (LCSW)—founder of the Veterinary Social Work program and certification through the University of Tennessee. In addition to working in the field of social work for more than 20 years, Dr Strand has many additional realms of expertise and certifications to support the myriad of initiatives that she has helped to create through University of Tennessee. In addition to the VSW Program, there are other certification programs, educational opportunities to support mental health and wellbeing, and peer mental health support platforms have been developed under Dr Strand's expert and compassionate guidance.
- Dr Jeannine Moga (MA, MSW, LCSW, PhD), clinical psychologist and MSW/PhD, founded the wellbeing program at U of Minnesota's veterinary teaching college, supported the efforts for ten years, and then developed and led Family & Community Services at NCSU's Veterinary Hospital. She is currently supporting wellbeing CE/training conversations via virtual learning (in person in the future!) for VetGirl and VEG. What a dynamo!
- Makenzie Peterson (MSc, completing her DSW in 2022), served as the first Wellbeing Program director at the Cornell University College of Vet Med 2018–2020. As of May 2020, she shifted into the role of the first director of wellbeing for the AAVMC. Per the AAVMC website, she will develop and implement specific programs in collaboration with the mental health professionals that are already working at many veterinary schools. It is likely that she will have a significant and positive impact on the discussions around curriculum development and the evolution of veterinary education to support the health of the individuals and of veterinary communities at large.
- Jen Brandt (LISW-S, PhD) is the AVMA director of Member Wellness and Diversity Initiatives since 2017. Dr Brandt is energetically contributing to the cultural shift in our profession to one of holistic wellbeing. There are a wide variety of programs that she has spear-headed through the AVMA that promote cultural humility, wellbeing in the workplace, and increased

mental health and wellbeing of individual associates. Get educated and get empowered! (www.avma.org for wellbeing resources and www.axon.avma .org for digital education).

DECREASING STIGMA AROUND MENTAL HEALTH AND SUBSTANCE ABUSE DISORDER/ALCOHOLISM

Talk about it! Knowing that veterinary medicine is an emotionally distressing profession based on research endeavors, feedback from veterinary professionals in preparation for this book, and personal experience, we all have a responsibility to decrease the persistent stigma around stress and mental health. It is critical that we "see each other" and give permission to ourselves and one another to acknowledge and express our feelings. This is an integral part of the "common humanity" portion of compassion, where we recognize that there is increased suffering when we feel isolated or that we are "the only one" coping with distress and difficult emotions. Practices should support regular dialogue and CE around stress and mental health issues in the veterinary environment.

It may be helpful as well to invite mental health professionals to speak at practice meetings, to contract with a veterinary-centric employee assistance program (EAP), and to encourage (and possibly incentivize) employees to participate in wellbeing CE and certifications. Access to mental healthcare overall also needs to be more flexible and inexpensive than it is currently. It is vital that everyone in the veterinary environment is aware of what mental health resources are available to them which might include EAP, tele-behavioral health options (e.g., E-counseling.com, 7cups .com, betterhelp.com), peer counseling, veterinary social work support, veterinary organization wellbeing resources, and virtual support through hotlines and web-based wellbeing resources (see Appendix 1 for more wellbeing resources).

SUICIDE AWARENESS AND SUPPORT

- Veterinary social workers and mental health professionals are increasingly on staff in veterinary schools and teaching hospitals and associated with the private veterinary organizations and individual practices
- There has been a marked increase in awareness and support of individuals struggling with mental health challenges and suicidal ideation in veterinary medicine in the last decade as evidenced by research endeavors and published articles, organizations created for support, and availability of trainings to increase accurate knowledge and decrease stigma in our community at large.

A sampling of initiatives created to increase awareness and support while decreasing the stigma:

Not One More Vet (www.nomv.org) and Facebook community for both veterinarians and veterinary technicians/parastaff. It was founded in 2014 following the suicide of the world-renowned veterinarian, Dr Sophia Yin. This

is a peer-to-peer online veterinary support group and has grown to support
a variety of resources dedicated around some of the toughest concerns we
navigate in veterinary medicine (e.g., cyberbullying).

Vets4Vets® is part of the VIN Foundation (VINFoundation.org) which is
a non-profit organization related to the Veterinary Information Network
(VIN).

Vets4Vets® is a free resource that is available to all veterinarians, not
just VIN members. These veterinary colleagues are prepared to support
both individuals and teams that need support around a wide variety of
mental health challenges. They also offer remote mentoring, peer support
groups, support for vet professionals in recovery, and collegial support for
those diagnosed with cancer.

Veterinary Confessionals Project (www.veterinaryconfessionalsproject
.com), founded by Dr Hilal Dogan in 2015 (she is currently the director of
wellbeing for the Veterinary Emergency Group).

Veterinary Interactive Screening Program (VISP) (www.visp.caresforyou
.org)

Created to support the mental health of veterinary professionals. This
is a partnership between University of Tennessee College of Veterinary
Medicine's Veterinary Social Work (VSW) program, NOMV, and the
American Foundation for Suicide Prevention (AFSP). As stated on their
website, the VISP is not intended to be either a crisis intervention or mental
health counseling site. Rather VISP is a mental health screening and peer
support service (completely voluntary and anonymous).

MINDFULNESS-BASED STRESS REDUCTION (MBSR)

Currently used in many human medical schools to support student mental health and
building of resiliency, MBSR is in its early days of finding its way into the veterinary
medicine community (classrooms and practices). The eight-week course is offered
in many spaces and ways but importantly is also now being offered through U of
Tennessee's VSW program. Here is a bit more about MBSR from the VSW website:

Mindfulness-Based Stress Reduction (MBSR) was founded by John Kabat-Zinn at the
University of Massachusetts Medical Center in 1979. The program was designed to
help human patients who were not responding to other forms of medical treatment.
MBSR has now spread to multiple populations including health professionals and
medical and nursing students as well as in multiple settings including work places,
educational settings and even prisons. UT VSW is the first to offer this class at a veterinary teaching institution.

Two decades of published research indicates that the majority of people who complete MBSR courses report:
Lasting decreases in physical and psychological symptoms

- An increased ability to relax.
- Reductions in pain levels and an enhanced ability to cope with pain that may
  not go away.

- Greater energy and enthusiasm for life.
- Improved self-esteem.
- An ability to cope more effectively with both short- and long-term stressful situations.

(http://vetscocialwork.utk.edu/mindfulness-based-stress-reduction)

BE THE CHANGE YOU WANT TO SEE IN THE WORLD

An apt way to bring this project of compassion and of hope to a close, I want to share some words of vision and of inspiration from veterinary colleagues. I asked these friends for their unique take on the question: "What are you excited about for the veterinary profession?"

MARIE HOLOWAYCHUK (DVM, DAVECC, SINGLE MOM, PIONEER IN VETERINARY WELLBEING SUPPORT/RESOURCES):

So much has changed in the veterinary world in just the last five years when it comes to mental and physical health and wellbeing. We seemed to have reached a peak of understanding when it comes to the struggles experienced by those who work in the veterinary profession. We know that our work comes with some unique stressors, that suicide is higher among veterinary professionals than that of the general population, and that there are many factors involved including access to controlled substances and mental illness. We also know that stigma persists among veterinarians and likely prevents individuals from seeking support.

However, in gaining this knowledge, we have also become more willing to share our experiences and speak openly about the issues. With more open discussions comes less stigma around mental illness and a normalization of seeking help.

Even more encouraging is a shift in focus away from the problems faced in the veterinary industry toward the solutions. We know that if people can develop tools and skills to boost resilience, not only can they foster wellbeing, but they are more likely to stay in the profession long term.

Several organizations, companies, practices, and individuals are dedicating time, energy, and resources toward developing and sharing these resources and supporting veterinary professionals. There is now a plethora of continuing education offerings, certificate programs, and coaching opportunities designed to share information related to effective team communication, conflict resolution, brave workspaces, diversity and inclusion, healthy boundaries, mental health first aid, and mindfulness, just to name a few. Whereas just a decade ago very few people were talking about workplace or professional wellbeing, now entire conference streams are dedicated to these topics.

I look forward to seeing how wellbeing continues to take shape within the veterinary profession and among veterinary professionals. I am thrilled to be a part of this journey and feel so fortunate to have inspired some people along the way.

Marie K. Holowaychuk (February 23, 2021)
(www.marieholowaychuk.com)

ADDIE REINHARD (DVM, SA GP, MS IN EDUCATION, RESEARCHER AND PROGRAM CREATOR):

I am excited to see that the veterinary profession is beginning to recognize and talk about the issues that we are facing. While we have known about the mental health issues among veterinary professionals for many years, historically, there have been very few targeted efforts to improve the mental health and wellbeing among veterinarians. Over the past five years, I have seen a growing number of initiatives that are attempting to improve the wellbeing of veterinarians. Also, veterinary colleges are making more of an effort to provide training in wellbeing to their veterinary students.

As we go forward, the more we talk about these issues as a group in a productive way, the more we will reduce the stigma in help-seeking behavior. I am looking forward to seeing what the next 10 years bring as I feel like we have the momentum to really make a positive impact! I think what will be key in moving forward is to ensure that we are communicating with each other as to what works and what doesn't work and ensure that our initiatives are grounded in research and best practices. By building the community of veterinary supporters, we will make the largest impact on the veterinary profession.

Brief bio: Addie earned her veterinary degree from the University of Tennessee in 2015. After practicing for several years in small-animal GP, she realized both that peer–mentorship was lacking for recent graduates and that she was passionate about pursuing further education to better inform herself how to approach this issue. She currently is a graduate student in the Department of Community and Leadership Development at University of Kentucky, researching the intersection of community development and veterinary wellbeing. Her overarching goal is to create a supportive community among veterinarians and empower veterinary professionals through self-care and leadership training.

AMY NEWFIELD (LVT, VTS-ECC, EDUCATOR, AUTHOR):

I have always had a love of unicorns. In the 1980s I was a unicorn junkie. To me the unicorn represented all things that were perfect, unique, and magical. When I was a child, I did my fair share of searching for unicorns which always resulted in me being disappointed. I always knew I was going to help animals and my love of all creatures, including the mythical ones, found me with a career in veterinary medicine as a credentialed veterinary technician.

I've been in the veterinary profession for over 20 years and have moved through small, single-owner veterinary practices to medium and eventually large corporate-owned practices. My most recent work has found me standing on a chair preaching about team culture and developing leaders. This comes from years of changing toxic workplace environments to healthier, happier ones. Eventually I took my love of unicorns, my passion for creating a healthier workplace environment and my desire to help leaders not struggle so much in roles they were provided no training for, and I wrote a book aimed at tackling all these issues.

I stand on the chair of team culture and leadership development because too many of our veterinary colleagues work in toxic, unhealthy hospitals. Gossiping, negativity, bullying, racism, and hazing run rampant in hospitals. They cause burnout and even suicide in our profession. Today I strive to create unicorns for the profession in every aspect that I am involved in.

I am hopeful for the profession because I see more unicorns than I have ever before. They are galloping beautifully alongside me with screams of healthier cultures in our hospitals and a focus on wellbeing. The best thing about this time now is you can have your voice heard through social media, podcasts, blogs, formal continuing education, or even by publishing a book with a unicorn on the cover. It will take an army of unicorns to change the veterinary profession, but that army starts with just one. #BeAUnicorn

Brief bio: With over 20 years in the industry, Amy brings her passion and love for education to VETGirl. After working in general practice, Amy found her passion in emergency medicine, and in 2003, she became a veterinary technician specialist in Emergency and Critical Care. She has held several board positions in the Academy of Veterinary Emergency & Critical Care Technicians, including president. Amy is well-published in over 15 subjects, is an international speaker, has received numerous awards, including two Speaker of the Year awards, and is highly involved in her community. The author of the book *Oops, I Became a Manager: Managing the Veterinary Team by Finding Unicorns*, Amy lives in Massachusetts with her wonderful furry kids, where you can find her eating chocolate, running in the woods, competing in agility, and diving in the ocean.

TIMBRALA MARSHALL (DVM, SA GP, MOM, VCA EI & DIRECTOR):

My name is Timbrala Marshall, and I am a Black, Female, Veterinarian in the Southern Region of the United States of America.

I want to point out several key pieces in that sentence: Black, Female, Veterinarian, South, USA.

All these words paint my story and provide context for my journey, but they do not define the legacy that I will leave.

Growing up in Selma, AL was a joy. I had parents who were committed to my success and family and friends who surrounded me with love. I was protected from the world that I know today, and I thank my parents for preserving my peace for as long as possible. This allowed me to focus on building my confidence, strengthening my voice, sharpening my knowledge, and developing my passion. At the age of eight, I was recorded by my late Uncle, saying, "When I grow up, I want to be a veterinarian". I said this boldly and without hesitation. I had never met a veterinarian. I had never stepped foot inside of an Animal Hospital. However, something inside of me determined that this would be my future.

Thirteen years later, I graduated from Tuskegee University with my BS in Animal and Poultry Sciences and entered Michigan State College of Veterinarian Medicine. My first day as a veterinary student, I experienced a defining moment—one that changed the trajectory of my life. For the first time ever, I was the only black person in the room, and I felt it in every way possible. Diversity was now my mission. As a young student, I charted the Michigan State CVM Chapter of VOICE (Veterinarians as One Inclusive Community for Empowerment). I am proud to see the work that continues at MSU CVM through their office of Diversity and Inclusion.

Today, I am the director of Veterinary Equity, Inclusion, and Diversity Programs for VCA Animal Hospitals. I stand with many others within our profession who are champions for change. My position allows me to work with underrepresented high school and college students through the creation of programs designed to give them exposure and experience in the veterinary industry. I also will help to ensure that associates from diverse backgrounds have the opportunity for promotion within the company into leadership roles. Our company is also committed to creating needs-based veterinary hospitals in underserved communities, and I am proud to be able to assist in these projects as well.

I am hopeful for the future of veterinary medicine and for the world. The younger generations are eager for change and demand equality. I firmly believe that each new day is a chance to change the future for the better, and my life's passion is to wake every morning with that goal in mind.

ANNE QUAIN (DVM, AUSTRALIA—UNIVERSITY, SA GP, EDUCATOR):

I am currently teaching DVM students, completing a PhD on ethically challenging situations encountered by veterinarians, animal health technicians, and veterinary nurses and locuming in companion animal practice. I contribute to a number of committees including the Australian Veterinary Association, the Humane Veterinary Medical Association's Leadership Council, and the Australian and New Zealand College of Veterinary Scientist's Animal Welfare and Ethics chapter.

Internally, my career trajectory has been a constant battle between curiosity and anxiety. When curiosity wins, I experience flow, I slow down, I ask more questions, I feel empowered, time expands. Everything is interesting. But at times, anxiety overwhelms curiosity, and then I feel like I am running through mud. Everything is difficult.

I've battled and will continue to battle with a relentless self critic (sometimes, they help me do my job well). And I feel like so many team members I work with are subject to the same harsh self-criticism. I am learning the things that feed my curiosity: making sure my basic needs are attended to (energy and hydration); stories (whether it's a client's narrative about their relationship with an animal or a great journal article or a book, movie, or play); or coaxing myself with small wins (how can I make today 1 per cent better?).

One thing that gives me hope is the concept of One Welfare. Somewhere along the line, I feel that thinking about animal welfare, human wellbeing, and environmental sustainability became disconnected—we forgot that we are all beings in a living world that sustains us. The concept that we need to address the welfare and wellbeing of animals, humans, and the environment makes absolute sense and has become a matter of urgency. In the context of the wellbeing conversation, its about recognizing that if veterinary team members have poor welfare, we cannot attend to our patients to the best of our ability. When we are putting out fires and trying to recover from exhaustion, we don't have the energy or bandwidth to turn our incredible knowledge and skill sets to address existential threats like climate change. With One Welfare I see a recognition that the needs of living organisms are interconnected, and we need to be mindful of systems and ecosystems.

Another thing that gives me hope is that finally, the discussion about wellbeing has zoomed out. The discussion increasingly recognizes that poor wellbeing of veterinary team members doesn't occur in isolation. We work in complex context and systems. Ethical fatigue and moral injury occur because of external constraints which need to be addressed at the level of systems. Our scholarly literature has shifted to reflect the knowledge that burnout does not signify moral weakness or resilience deficiency of an individual. It's a marker of a system issue.

It is great to see the shift from discussions in scholarly and gray literature around wellbeing of veterinarians to wellbeing of veterinary teams. This is so important. Veterinarians do not have a monopoly on mental health issues. Its also great to see examples of veterinary teams connecting with people from other professions: health care teams, law enforcement, hospitality workers. The us/them mentality serves no one. Yes, veterinary team members are exposed to unique stressors. But people in other roles are exposed to unique stressors too. I'm glad to see the conversation shifting from "we have the HIGHEST suicide rate" to "what are you guys doing to look after your people that works?".

"Quick and dirty biography": Dr Anne Quain lectures at the Sydney School of Veterinary Science and is a companion animal veterinarian. She completed a Masters in small animal medicine and surgery through Murdoch University, is a member by examination in the animal welfare chapter of the Australian and New Zealand College of Veterinary Scientists, and is a diplomate of the European College of Animal Welfare and Behaviour Medicine in Animal Welfare Science, Ethics and Law.

Dr Quain coauthored *Veterinary Ethics: Navigating Tough Cases* with Dr Siobhan Mullan, and coedited the *Vet Cookbook*. She is a member of the AVA's New South Wales Executive Committee, the Animal Welfare Advisory Council, and the Humane Society Veterinary Medical Association leadership council.

CLAY CHILCOAT (DVM—USA— RESEARCH, EQUINE, MILITARY)

Talking. The best thing going on right now in our profession is that we are starting to communicate among ourselves about our emotional and psychological stress. As a mentor to young veterinarians, I am focusing on openly sharing my problems and encouraging them to do the same. Regardless of the cause—be it the personal traits that draw us to the profession, the selection factors to get into vet school, the "suck it up" mentality of some veterinary faculty, the relative physical and social isolation from other veterinarians, ad nauseam—I believe veterinarians tend to hide their struggles from the world, and by doing so, we cannot see that others have the same problems. By sharing our issues, we do more than just permit ourselves to be helped and understood, but also, we demonstrate to others that they are not alone in their struggles. The attention that veterinary emotional and psychological issues are receiving both within and outside the veterinary community is the catalyst to getting the conversations flowing. I believe it is crucial for this effort to be led by more senior veterinarians. While many of us may have achieved the resiliency over time that our younger colleagues are still searching for, it is invaluable for us to demonstrate that we had (and still cope with) the same stresses which they face. Basically, if I can make it through then you can too! We have to keep talking.

Brief bio: Dr Clay Chilcoat is an Army Veterinary Corps officer and proud "vet brat" as the son of a companion animal practitioner. His veterinary career has meandered from equine clinician to research scientist and now to public health. He lives in Virginia with his wife and horses, where he pursues dressage and amateur fence repair.

LINDA FINEMAN, DVM, ACVIM (ONCOLOGY), CHIEF EXECUTIVE OFFICER, ACVIM AND ERIKA PICCIOLO, LVT, MS, EDUCATOR, ACVIM:

The last year has been one of the most painful and jarring times many of us have ever experienced. Yet with the massive disruption caused by the global pandemic comes opportunity for growth and change. The American College of Veterinary Internal Medicine (ACVIM) has been built upon a foundation of community. Finding ways to foster that sense of connection during a prolonged period of isolation has been one of our biggest challenges and will continue to be an area of focus as we move gradually into whatever our "new normal" is.

As a large organization with members all over the world, the annual ACVIM Forum has been a primary way for our members [to] connect with colleagues and friends, reinforcing our bonds as individuals who have a shared set of experiences. For 2020, cancellation of this event meant a rapid transition to a virtual event, with little time to create nuanced opportunities for social networking and conversations. Looking ahead, we see the opportunity in hybrid continuing education events, allowing us to harness the power of online sessions for didactics as an adjunct to more interactive live sessions. We are reimagining all of our educational offerings by creatively thinking about how and when content is delivered and reflecting our shared understanding of the importance of human connection. By thinking intentionally about the separate but related goals of education and connection, we have the opportunity to extend our reach through virtually delivered educational content available before, after, or in place of, a live event. Hybrid courses combining recorded content and live virtual components provide opportunities to learn in a self-paced environment, while also coming together for real-time Q&A with subject matter experts. As we create a new vision for the near-term future, we expect augmented reality and a choose-your-own-adventure structure to be a means of creating robust virtual learning experiences for technical, medical, and professional skills.

While the effects of the pandemic provide impetus for a revamped and reinvigorated conference experience, we also envision a much more intentional focus on interactive educational opportunities, social and professional networking, relationship-building, and connection outside of convention-like settings. We are realigning our work with our members' needs in a world that has fundamentally changed, with no end in sight to the pace of rapid innovation and change. Our members are looking for, and finding new ways, to leverage their professional skills and subject matter expertise and to thrive in our profession. As leaders in veterinary specialty medicine, the ACVIM must also find new ways to collaborate and drive change. Working hand-in-hand with other veterinary medical associations, we strive to bring new ideas to our members and to the profession.

The future of the ACVIM looks bright: we have a passionate group of members, many of whom are dedicated volunteers, supported by our mission-driven staff. Together, we all feel a deep sense of commitment to the profession and to our mission of improving lives of animals and people globally. We've been inspired by the role our members played in advising other epidemiologists and reporters on infectious diseases, sharing their equipment and expertise, and working long, stressful hours to help people and their animals. We are energized by how the ACVIM and our members will be leaders in the future of veterinary medicine as we tackle the difficult issues of mental health, wellbeing, lack of diversity and sustainability—all critical to our profession.

BETSY CHARLES (DVM, RADIOLOGIST, EQUINE, ACADEMIA—WSU, VLE, ENTREPRENEUR AND EDUCATOR):

I am emerging from 2 years of reorienting myself to a new normal, a normal that doesn't include my favorite person in the world. Getting to walk through the valley of the shadow of death with my beloved Drake, and then ushering him into the arms of his Creator, changed me dramatically. It transformed how I show up in the world. I am definitely still figuring out who the new normal Betsy is and what she is going to bring to the world, but the pep-stepping, get 'er done, workhorse with some sharp edges Betsy is, for sure, not the same. Now, don't get me wrong: I will always be a pep-stepping, get 'er done workhorse—it's who I was created to be, I think. But I now take the time to saunter every once in a while. I am definitely more interested in being fully present wherever I find myself each day. And, my edges are a little smoother.

I am also very, very aware of how the things I have spent my career working on (the Veterinary Leadership Institute, rethinking how we educate veterinary students, challenging preconceived ideas of what leadership looks like in vet med, etc.) have been critical to my ability to rise from the ashes of losing my best friend. I am learning how to prioritize the things that matter, which means I need to practice the skill of saying no and setting healthy boundaries—not an easy thing for me. I have had the opportunity to practice what I preach more than ever and it is these experiences that cause me to be super excited about having a voice in the wellbeing conversation within veterinary medicine.

DR MARCI KIRK:

I feel like my story was pretty typical. I wanted to become a veterinarian for as long as I can remember and spent my entire life pursuing this goal. I had great visions of what that would be like, sure, I knew there would be challenges, but there would also be moments where I saved a life and have a positive impact on someone's day. My transition from student to veterinarian is not what I would call graceful. I struggled a lot with Imposter Syndrome as well as very Black and White thinking (I'm either a great vet or I get a board complaint—there was no middle ground in my mind). This led to a tremendous amount of angst at starting each working day and continued to bleed over into my non-work life. My thoughts revolved around one phrase: "Maybe I wasn't meant to be a veterinarian after all".

It turns out, I had a very narrow view of what being a veterinarian meant. The DVM or VMD degree unlocks so many opportunities. We just have to be brave enough to take the next step toward them. When I left private practice to join the AVMA, I never thought I would go back to clinical medicine, but after some time away, and a lot of perspective, I realized how much I missed it. So, I've created a working life that works for me. My full-time position is as the assistant director for Recent Graduate Initiatives with the AVMA. Here I help create programs that offer guidance for that transition from student to veterinarian. Through this work I have gotten to meet so many different colleagues from across all areas of the profession and feel constantly inspired by what they are accomplishing. I also get to work closely with Dr Jen Brandt PhD, LCSW, AVMA's director of AVMA Wellbeing Initiatives. I've started dipping my toes back in clinical medicine through several relief shifts per year. I've found this combination really works for me. I believe we have some of the brightest and best people in the veterinary profession. All it takes is a little inspiration and some creativity and you can find your place in veterinary medicine. I truly believe everyone deserves to be here, and we are better for you being here.

Brief bio: Dr Marci Kirk is a 2011 graduate from the University of Illinois College of Veterinary Medicine. She practiced small-animal medicine for five years before joining the staff at the AVMA where she is currently the assistant director for Recent Graduate Initiatives. Outside of work, Dr Kirk is an avid runner and enjoys traveling with her husband, especially to Walt Disney World, usually for a race weekend. She also shares her home and most of her social media posts with her two dogs, Charlie, the best good boy and Bucky, the aspiring good boy.

Parting Thought: Let's Live into the Spirit of Ubuntu

As shared by former American President Barack Obama, when he made a 2018 speech honoring Nelson Mandela:

> There is a word in South Africa, "Ubuntu", that describes his greatest gift: his recognition that we are all bound together in ways that can be invisible to the eye; that there is a oneness to humanity; that we achieve ourselves by sharing ourselves with others, and caring for those around us.

Appendix 1:
Additional Resources

MENTAL HEALTH AND SUBSTANCE ABUSE DISORDER RESOURCES

UNITED STATES

- Mental Health.gov (www.mentalhealth.gov)
- National Alliance on Mental Illness (www.nami.org/Find-Support/Nami -HelpLine)
- 9-1-1—if someone may be in immediate danger to themselves or others.
- 9-8-8—suicide prevention hotline coming in 2022.
- National Suicide Prevention Lifeline 1-800-273-TALK (8255).
- Crisis Text Line
  - Text "HOME" to 741 741, and a trained counselor will respond.
  - Type "mobile crisis" into search engine for assistance with mental health crisis in virtually every community.
- IMAlive (www.imalive.com)—provides a free, confidential, and secure online chat service. All chats are answered by trained volunteers.
- National Domestic Violence Hotline (800) 799-7233.
- Rape, Abuse, and Incest National Network (RAINN)—(800) 656-HOPE (4673) or hotline.rain.org
- National Institute on Drug Abuse Hotline (800) 662-4357.
- Better Help (www.betterhelp.com)—the world's largest e-counseling platform.
- Self-Harm Hotline—1-800-DONT CUT (366-8288).
- LGBT National Help Center—1-888-843-4564.
- Trans Lifeline—www.translifeline.org or 1-877-565-8860.
- Veterans Crisis Line—www.veteranscrisisline.net

WEB-BASED MENTAL HEALTH SUPPORTS

7 cups—www.7cups.com
e-Counseling.com—www.e-counseling.com
www.suicidestop.com (international resources here).

UNITED KINGDOM

- Vetlife (www.vetlife.org.uk)
- 24/7 Helpline—116 123 (UK and ROI)

- Samaritans.org
- YourLifeCounts.org

AUSTRALIA

- www.Lifeline.org or 1-300-13-11-14
- YourLifeCounts.org/find-help
- www.BeyondBlue.org.au
- This Way Up (https://thiswayup.org.au)

CANADA

- Canadian Mental Health Association (https://cmha.ca)
- CrisisServicesCanada.ca/en
- Canadian Association for Suicide Prevention (www.suicideprevention.ca/need-help/) or 1-833-456-4566.
- YourLifeCounts.org/find-help/

VETERINARY ORGANIZATION WELLBEING SUPPORTS

- Australian Veterinary Association: www.ava.com.au/member-services/vethealth
- American Veterinary Medical Association: www.avma.org/resources-tools/wellbeing
- Canadian Veterinary Medical Association: www.canadianveterinarians.net/veterinarianwellness
- National Association of Veterinary Technicians in America: www.navta.net/page/Wellbeing

OTHER VETERINARY WELLBEING RESOURCES

- IMatter: https://i-matter.ca
- Mighty Vet: https://mightyvet.org
- The Banfield Exchange: www.banfield.com/exchange/life-practice/personal-health-wellbeing
- VETgirl: https://vetgirlontherun.com/tag/wellness
- VIN Vets4Vets: https://vinfoundation.org/resources/
- MVH4You: https://mvh4you.com

FINANCIAL PLANNING RESOURCES

Student Debt Help: VIN Foundation Student Debt Center (vinfoundation.org/resources/student-debt center)

FINANCIAL PLANNING

- National Association of Personal Financial Advisors (NAPFA.o rg)—fee-based
- Veterinarian Financial Advisor Network (veterinarianfinancialadvisorn etwork.com)

PRO BONO FINANCIAL COUNSELING

- NAPFA Foundation (napfafoundation.org)
- Financial Planning Association (onefpa.org/advocacy/Pages/pro_bono _financial_planning. aspx)
- National Foundation for Credit Counseling (nfcc.org)

AVMA FINANCIAL TOOLS (ON THEIR WEBSITE)

1. Business 101 for Non-Economists (whether you are running a veterinary practice or applying business knowledge to your own professional income); building confidence by increasing exposure to terms and to articles/writing; understanding the pillars of business simply and profit-building.
2. Personal Financial Planning Tool (PFP Tool)
3. Interviews and helpful tips supporting financial wellbeing, budgeting, loan repayment strategies, debt management, calculating for large expenditures (such as buying a car or a home), and retirement savings concepts.

SMARTVET FINANCIAL ADVISING

Financial Advisors specific to veterinary professionals are rare. A pioneer in this realm is the SmartVet Financial Advising group based out of Tampa, FL. Cofounders Tom Seeko and CJ Harnett have been involved in wealth-management and financial advising for many years. Their team's focus in the last three years, however, has been to better understand the unique stressors and financial dynamics for veterinary professionals. In addition to providing financial coaching and advising, their desire to support financial wellbeing for veterinarians by providing knowledge has also resulted in the development of CE and financial guidance specific to veterinary professionals starting in veterinary school. (Discuss what they are currently offering with regards to their CE endeavors, podcasts, and online tools).

Appendix 2:
Uncle Mikey's Maxims

As compiled by Dr Michael Schaer, DACVIM, DACVECC, professor emeritus and adjunct professor of ER and Critical Care Medicine at University Florida College of Veterinary Medicine.

1. Treat for the treatable.
2. Assumptions lead to trouble; therefore, don't assume.
3. Always interpret clinical information within the context of the patient's presentation.
4. Avoid tunnel vision.
5. Treat your patient, not just its disease.
6. Avoid over-medicating.
7. Be honest with yourself.
8. Don't postpone today's urgencies until tomorrow.
9. Think that common things occur commonly.
10. Look closely at your patient; it will usually tell you what's wrong.
11. Never let your patient die without the benefit of the silver bullet.
12. When you hear hoofbeats, look for horses, but don't forget about the zebras.
13. Never sell the basics short—they are still the best buy in town.
14. If you don't think it, you won't find it.
15. Never let a biological specimen go to waste.
16. Disaster lurks whenever a patient's problem is routine.
17. If it's not getting worse, give it a chance to get better.
18. Don't stray too far from the patient—the diagnosis will appear eventually.
19. Don't give your patient a disease it doesn't deserve to have.
20. Don't let technology make you decerebrate.
21. The necropsy is the clinician's trial by jury.
22. The wisdom of experience should never be ignored.
23. The diagnostician should always ask him/herself these two questions: "Where am I with this patient?" and "Where am I going?"
24. If the patient isn't going where you expect it to be going, then go back to square one.
25. In order to successfully treat a cat, you must think like a cat.
26. Avoid the pitfalls of the red herring.
27. If they can't afford a caddy, then offer them a chevy.
28. Know thy patient.
29. Nobody wants to pay for a big bill and a dead animal.

30. What matters is not so much what you say to a concerned client, but how they perceive what you said.
31. Diagnostic cloudiness will soon be replaced by clear skies.
32. Better that the dying patient expire in the hospital than during the car ride home.
33. You must have cognition to be a competent diagnostician.
34. To prognose you must first be able to diagnose.
35. The "toaster effect" will occur when your patient reaches the turning point toward recovery.
36. To cut is to see; to see is to do; to do is to cure.

Recommended Readings

BOOKS

1. *Clin Life-21: The Veterinary Guidebook* by Quincy Hawley, Caitlin Keat, Renee Michel, Alyssa Mages to help you build clinical skills and cultivate wellbeing habits.
2. *Thrive* by Arianna Huffington
3. *Resilient* by Rick Hanson, PhD
4. *Grit* by Angela Duckworth
5. *The Resilient Practitioner* by Thomas M. Skovhold and Michelle Trotter-Mathison
6. *The Compassion Fatigue Workbook* by Francoise Mathieu
7. *The Gifts of Imperfection* and *Dare to Lead* by Brené Brown
8. *Coping with Stress and Burnout as a Veterinarian* by Nadine Hamilton
9. *Oops! I Became A Manager* by Amy Newfield
10. *Staying Sane in the Veterinary Profession* by Annette Docsway, DVM
11. *Veterinary Ethics: Navigating Tough Cases* by Siobhan Mullan & Anne Quain (Fawcett)
12. *Trauma Stewardship* by Laura van Dernoot Lipsky
13. *Communication Skills* by Monica and Suzanne Kurtz
14. *The Unspoken Life: Recognize your Passion, Embrace Imperfection, and Stay Connected* (2017) by Kimberly Pope-Robinson. Check out her 1 Life Connected Coloring Book and her website for other fun resources, such as her card decks for values and emotions. (https://1lifecc.com)
15. *Manifesto for a Moral Revolution* by Jacqueline Novogratz
16. *The Four Tendencies* by Gretchen Rubin
17. *The Passion Test: The Effortless Path to Discovering Your Life Purpose* by Janet Bray Attwood & Chris Attwood
18. *Vet School Survival Guide: Notes from a Back Row Student* (2017) and *Vet School Survival Guide II: Vet Med Spread* (2019) by Dean Scott.
19. *Perfectly Hidden Depression: How to Break Free from the Perfectionism That Masks Your Depression* (2019) by Margaret Robinson Rutherford
20. *Fierce Self-Compassion* (2021) by Kristen Neff

Facilitator's Guide

CHAPTER 1:

1. When you reflect on the historical and current challenges facing veterinary professionals, what are some of the top issues that come to mind right now?
2. What is the meaning of "veterinary medicine professional" to you?
3. The human–animal bond has evolved in the last 50 years and plays a significant role in current veterinary medicine. What are some of the challenges and some of the positive opportunities that you can think of about this important bond between animals and their guardians?
4. The veterinary profession offers a wide variety of environments where animal care can be applied. Consider the veterinary segments that you have personal experience with and share what you believe to be unique to those environments regarding patient care and potential stressors for the caregiving professionals.

CHAPTER 2:

1. Can you explain the difference between "rumination" and "reflection"?
2. There are many ways to consider the term "stress". Share some of the key takeaways for you after reading more about stress, eustress, and distress.
3. The difference between compassion fatigue and empathic distress is a vital one for caregivers. How are these mental health challenges similar and how do they differ? How do you think having this clearer understanding might positively benefit you or other veterinary professionals?
4. Burnout is commonly discussed as a caregiving concern or a likely sequela of being in a caregiving profession. Do you understand the differences between compassion fatigue, empathic distress, and burnout? Why do these differences matter, in your opinion?
5. Moral distress and ethical conflicts are relatively new terms in caregiving professions. After having the definitions clarified, could you offer some examples of each that you experienced or observed in veterinary practice?

CHAPTER 3:

1. In the formation of professional identity, there are four stages that were discussed and considered with regard to the unique challenges and opportunities. Please list the stages and share some of the issues that were raised or others that are valuable to consider in your experience that impact healthy culture, mental health, and professional development.

2. In your professional experience to date, what are some of the cultural and mental health challenges that you noted in different veterinary environments? Were there any unique strategies or initiatives implemented that positively impacted these challenges?

3. Trauma as a concept in caregiving environments and trauma-informed care were discussed in this chapter. What is your understanding now of trauma-informed care and associated language? How can these concepts be applied in the clinical environment, e.g., when speaking to clients?

4. Perfectionism and imposter syndrome are common concerns for high-achieving individuals, such as veterinary professionals. What are some ways that these tendencies can be self-defeating and how might they be "flipped around" to support growth and development?

5. Moral distress and ethical exhaustion in veterinary medicine were examined in more depth in this chapter. Please take time to review some examples from the chapter or identify others that you have experienced/witnessed and consider the emotional impact on yourself and your colleagues.

6. How would you now define substance use disorder and how can we support our community in addressing this stigmatized mental health challenge?

7. The issue of suicide in the veterinary profession is a significant pressing concern. What did you learn that you did not know already in this chapter? Do you have a better sense of the resources that are available to support colleagues that may need compassionate, informed care?

CHAPTER 4:

1. What are the three components that were discussed to support self-compassion? Can you give some examples for each component?

2. What are the eight realms in the "wellbeing wheel"? Can you provide a few examples for each realm?

3. What is meant by "sleep hygiene?" Why is healthy sleep necessary, and what are some ways that you can identify for yourself that can support better rest?

4. As you examine your own coping strategies, what is serving you and what may need to be reconsidered?

5. Boundaries are integral to self-care and to building resiliency. What are some boundaries that you can identify that you have already set and where might you benefit from boundary creation and communication?

6. What are some of the positive impacts of practicing mindfulness? What are some mindful practices that you feel may benefit and support you?

CHAPTER 5:

1. What is meant by "compassion satisfaction"? How can this counter-balance compassion fatigue?

2. Positive psychology was raised as an exciting and helpful framework for evaluation of our professional systems and strategies to support flourishing individuals and teams. How could you see this work impacting some of the professional concerns that you are aware of today?

3. Eudaimonia is likely a new term to add to your vocabulary. How would you describe this concept to a friend or family member? Why would the framework of eudaimonia be valuable when strategizing on how to create healthier, sustainable systems to support the veterinary professionals now and in the future?

CHAPTER 6:

1. This chapter was intended to share the bounty of resources that have been, or are being, developed to support veterinary professional wellbeing and development. What platforms or resources were new to you?

2. After reviewing the many ways that veterinary colleagues are supporting the robust conversations around building a compassionate, forward-thinking, caregiving community, could you see yourself wanting to get involved or to learn more about any of these initiatives?

3. What is missing in the offerings shared and is there something that you uniquely can create and contribute?

Acknowledgments

The contents of this book have been developing inside my brain and heart for years prior to finding the courage to say "yes" to the actual writing. There have been few resources for those of us in veterinary medicine who seek to put names to what we are feeling and experiencing. As a result, I sense that there has been, and is, so much unnecessary suffering and concurrent creation of false narratives (there is something "wrong" with us to be feeling the way that do).

As I sought conversations and education from veterinary colleagues and other caregivers in the world to better frame the experiences I had had as a veterinarian, it was uplifting and restorative to realize that we were all trying to figure this out: how can we provide sustained, compassionate care for others and not lose ourselves in the process? In the midst of these years of wellbeing practices and CE, of "heart-storming" discussions with friends and family, and introspective work, the ingredients for this book started to come together, mix, and simmer—like an authentic curry. Each and every conversation and clarifying experience was invaluable along the way. With that said, this is the moment where I have the opportunity to honor and express immense gratitude for:

- First and foremost, the yang to my yin, Matty D—aka, my husband. He has been unfailingly supportive in giving me the time, space, and support in all the ways needed to keep me focused on the work needed to bring this book to life.
- My parents—Dad for always believing in my capacity to do anything I put my mind to and for providing me the means to pursue my professional dreams. Mom for infusing me with a respect and feeling of connection to all living things, particularly animals.
- My dear friend, mentor, and amazing life coach of nearly six years who is an inspirational wonder woman in her own right, Victoria Travis. Her wisdom and compassionate guidance provided me with the courage, perspective, and knowledge to believe in my worth in the world and live a life that I love.
- My "work wife" and close friend, Jan Massaro, for always believing wholeheartedly in me and that my voice mattered in supporting the veterinary wellbeing conversations. She never wavered in her conviction that I was the one that needed to write this book.
- Dr Seth Vredenberg, my friend and coconspirator in taking this book project from our animated, inspirational conversations to finding an amazing advocate in Alice Oven with Taylor & Francis Publishing. Seth has been, and is, another amazing veterinarian who is committed to all veterinary professionals having fulfilled and healthy lives.

- The generosity of Dr Dean Scott for sharing his artwork and fabulous veterinary-centric cartoons which I honestly could not imagine this book being the same without!
- The book cover was a wonderful synergy of being the photographic embodiment of the human–animal bond on a life journey that came from my dear friend, Pam Stevenson, and from Janelle Straszheim and her horse, Samie, who brought it to life on a trail in Poolesville, Maryland, very close to where I grew up and used to ride with Janelle.
- The kindred spirts and creative, collaborative energy of Vanessa Rohlf and Alice Oven that have been a part of this journey from Day 1. Together, we shared the conviction that together we could bring awareness, knowledge, and inspiration to the veterinary professional community that we each care so much about.
- The incredibly creative and energizing conversations I had with Anne Quain, Karen Fine, Alyssa Mages, Marie Holowaychuk, and Meredith Kennedy, who each kept me motivated and committed to this work. I have learned so much from each of these amazing women and am inspired by the work that they are each doing in the veterinary wellbeing realm as well.
- Helen Chang with AuthorBridge for the valuable lessons in book creation and, more importantly, her coaching on relatable storytelling. She got me unstuck and moving forward with confidence and excitement when I got entangled in my perfectionist, imposter-syndrome thought processes.

And to so, so many friends and colleagues who, when I would discuss the book as a germ of an idea, were willing to speak with me and share their thoughts and life wisdom. Their stories, reflections, and energy are all part of the weave of this book's story. Immense thanks go to:

Addie Reinhard, Amber McAllister, Amy Newfield, Beth Davidow, Betsy Charles, Brian Stewart, Caleb Frankel, Casara Andre, Cindy Trice, Clay Chilcoat, Dani McVety-Leinen, David Bessler, Debbie Stoewen, Delian Nichols, Diana Barone, Dondrae Coble, Elizabeth Strand, Emily Weaver, Emma Whiston, Erika Picciolo, Francesca Tocco, Gary Hurlock, Holly Anne Stringfellow, Jean Sonnenfield, Jeannine Moga, Jennifer Lopez, Josh Vaisman, Julia Jones, Kate Brammer, Katie Ford, Ken Drobatz, Kimberly Pope-Robinson, Laura Buscher, Linda Fineman, Lisa Paull, Lori Harbert, Lucy Spelman, Lucy Squire, Lynn Roy, Marci Kirk, Mark Olcott, Michael Schaer, Mike Endrizzi, Nadine Hamilton, Pam Stevenson, Phil Richmond, Quincey Hawley, Rita Hanel, Sinead Greer-McAuley, Steve Baker, Steve Kochis, Steve Tutela, Susan Swendsen-Harris, Susan Hazel, Tamara Hancock, Teresa Lightfoot, Timbrala Marshall, Tosha Zimmerman, and Traci Delos.

Index

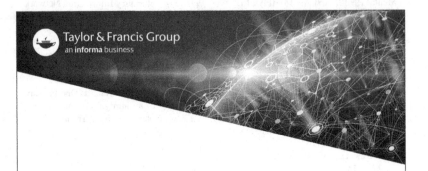